EXPERIMENTS IN GENERAL CHEMISTRY

Principles and Modern Applications

Thomas G. Greco • Lyman H. Rickard • Gerald S. Weiss
Millersville University

Petrucci • Harwood • Herring

Eighth Edition

Prentice
Hall

Upper Saddle River, NJ 07458

Editor in Chief: John Challice
Project Manager: Kristen Kaiser
Executive Managing Editor: Kathleen Schiaparelli
Assistant Managing Editor: Dinah Thong
Production Editor: Meaghan Forbes
Supplement Cover Manager: Paul Gourhan
Supplement Cover Designer: PM Workshop Inc.
Manufacturing Buyer: Alan Fischer

© 2002 by Prentice-Hall, Inc.
Upper Saddle River, NJ 07458

Printed in the United States of America

10 9 8 7 6 5 4 3 2

ISBN   0-13-017688-5

Prentice-Hall International (UK) Limited, London
Prentice-Hall of Australia Pty. Limited, Sydney
Prentice-Hall Canada, Inc., Toronto
Prentice-Hall Hispanoamericana, S.A., Mexico City
Prentice-Hall of India Private Limited, New Delhi
Pearson Education Asia Pte. Ltd., Singapore
Prentice-Hall of Japan, Inc., Tokyo
Editora Prentice-Hall do Brazil, Ltda., Rio de Janeiro

ACKNOWLEDGMENTS

Every book is the product of many individuals, and this one is no exception. Many improvements have resulted from the support and efforts of the Chemistry Department faculty and staff at Millersville University throughout the writing process. We are especially grateful to Cathy Greco, Lynette Rickard, Edward Rajaseelan, and Sandra Turchi for their contributions to this edition.

Gerald S. Weiss
Thomas G. Greco
Lyman H. Rickard

Preface

When studying beginning chemistry, students are introduced to many theoretical concepts based on conclusions drawn from years of accumulated observations. It is difficult for many of them to appreciate the importance of experimental observation and its relation to theory, especially when they have had limited practical experience. Accordingly, this laboratory manual is designed to provide the beginning chemistry student exposure to the basic techniques of laboratory work and the practical experience necessary to understand and appreciate better the general information presented in the text and lectures.

The experiments in this manual have been selected with a fourfold purpose:

1. To introduce the student to the fundamentals of physical observation and to the scientific method.

2. To acquaint the student with several basic quantitative techniques, including gravimetric and volumetric measurement, the care and use of laboratory equipment, and the safe use and handling of chemicals.

3. To provide empirical verification–that is, information based on observation or experience–of chemical theory in a qualitative and quantitative fashion.

4. To do the above in close coordination with the ideas and mathematical concepts presented in the eighth edition of *General Chemistry* by Ralph H. Petrucci and William S. Harwood. Designed to complement the Petrucci/Harwood text, this laboratory manual provides a unified experience in chemistry at the introductory level, both for chemistry majors and for students in closely allied fields.

Each experiment includes an introduction, experimental and waste disposal procedures, report tables for data and results, questions, a prelaboratory assignment, and space for sample calculations. There are three types of student disposal instructions:

1. *Recycle or reuse:* Wastes are collected in labeled bottles for either direct reuse or for further treatment prior to reuse.

2. *Laboratory treatment and disposal:* Students will neutralize acids and bases, reduce chemical activity, or filter wastes prior to disposal in a safe and appropriate manner.

3. *Disposal:* Wastes are collected in labeled glass disposal bottles, which should be segregated according to chemical compatibility. These will be stored for further treatment or volume reduction by laboratory personnel prior to or for direct disposal by a licensed waste disposal company.

The introduction contains enough of the related principles and learning objectives of the exercise to allow the student to complete the prelab assignment, answer the questions, and carry out the laboratory work. However, time should be allowed beforehand for the laboratory instructor to discuss and demonstrate specific techniques, safety precautions, common problems, and other aspects of that experiment.

Completion of the prelaboratory assignment requires that the student read the experiment and become familiar with the procedure before coming to the laboratory to do the experiment. In this way, students learn to prepare themselves properly for the experiment, make fewer mistakes, and become more efficient in utilization of available time. We believe this to be an essential part of the complete experiment.

Emphasis is placed on experimental precision and on obtaining accurate results. Many experiments involve the determination of some "unknown" quantity. This necessitates good technique and reproducible work. Thus a knowledge of precision and accuracy, which are discussed in the Introduction, is essential. Such knowledge is then used in the laboratory report that accompanies each experiment. The laboratory report often consists simply of the completion of the data tables, sample calculations, and answers to the questions that come with the experiment. The writing of a laboratory report is described in detail in the Introduction.

It should be emphasized that laboratory data should *not* be recorded originally on the enclosed sheets; rather, they should be recorded in a separate, bound laboratory notebook, which is the original record of all experimental observations and data. A detailed description of the guidelines for the laboratory notebook is also given in the Introduction. The reports are then written from data and the first draft of the answers to questions in the notebook.

Many more experiments are included in this manual than can be used in a normal one-year general chemistry laboratory schedule. For example, there is more than one experiment dealing with several topics such as gases, chemical reactions, equilibria, and acids and bases. With that consideration, however, the sequence of experiments follows closely the order of topics presented in the Petrucci/Harwood text. Experiments can thus be selected which provide application to each topic covered here. In this way, a laboratory program can be designed that follows closely and correlates well with the text. We believe such correlation to be essential to a well-designed, sound introductory chemistry course.

A number of descriptive experiments are included which we believe are unique in a laboratory manual at the introductory level. These include several experiments: 3 and 4 on chemical reactions; 6, "Identification of Common Chemicals"; 18, "Weak Acids and Bases and Their Salts"; 28, "A Penny's Worth of Chemistry"; and 31, "Free Radical Bromination of Organic Compounds." In addition, several traditional descriptive preparations of specific chemicals are included (Experiments 27, 29, and 30). Also, a condensed version of the traditional qualitative analysis scheme is provided that is meaningful and can be completed in about six laboratory periods. Certain other experiments require more time than is available in the normal two- to three-hour laboratory period. These experiments can either be extended over two periods or shortened to the time span available. The instructor's manual provided for this laboratory manual includes such specifics.

Many of the experiments are reduced in scale in an effort to minimize the amount of chemicals required, to increase safety, and to reduce the amount of waste chemicals. We have also included instructions regarding the proper disposal of waste chemicals for each experiment. A completely new set of prelab questions and many report questions have been included in this edition.

It is our conviction that time is valuable and too short in a two- or three-hour session to allow for more than an introductory investigation of most topics. However, a thorough treatment of fundamentals is possible. This is particularly true in a beginning course. The student will have greater opportunity in later courses to experience more advanced methods.

It is essential that students acquire and appreciate sound experimental technique early and learn to apply it with confidence. They will quickly discover that chemistry is still very much an experimental science.

G. S. W.
T. G. G.
L. H. R.

Contents

viii   Contents

Laboratory Safety Rules

In the home, the kitchen and bathroom are the sites of most accidents. In a school, the chemical laboratory poses similar hazards, and yet it can be no more dangerous than any other classroom if the following safety rules are always observed. Most of them are based on simple common sense.

1. Behave responsibly. The dangers of spilled acids and chemicals and broken glassware created by thoughtless actions are too great to be tolerated.

2. Wear approved eye protection *at all times* in the laboratory and in any area where chemicals are stored or handled. Such protection will protect you against impact and chemical splashes. Goggles are strongly recommended and may be required. The only exception is when explicit instructions to the contrary are given by your instructor.

    a. If you should get a chemical in your eye, first rinse with isotonic sterile solution, then wash with flowing water from a sink or fountain for at least 15 min. Get medical attention immediately.

    b. Do not wear contact lenses in the laboratory, even with safety goggles. Contact lenses prevent rinsing chemical splashes from the eye. Vapors in the laboratory (HCl, for example) dissolve in the liquids covering the eye and concentrate behind the lenses. "Soft" lenses are especially bad as chemicals dissolve in the lenses themselves and are released over several hours.

3. Do not perform any unauthorized experiments. This includes using only the quantities instructed, no more. Consult your instructor if you have any doubts about the instructions in the laboratory manual.

4. Do not smoke in the laboratory at any time. Smoking is not just an obvious fire hazard; it also draws chemicals in laboratory air (both as vapors and as dust) into the lungs.

5. In case of fire or accident, call the instructor at once. Note the location of fire extinguishers and safety showers now so that you can use them if needed.

    a. Wet towels can be used to smother small fires.

    b. In case of a chemical spill on your body or clothing, wash the affected area with large quantities of running water. Remove clothing that has been wet by chemicals to prevent further reaction with the skin.

6. Report all injuries to your instructor at once. Except for very superficial injuries, you will be required to get medical treatment for cuts, burns, or fume inhalation. (Your instructor will arrange for transportation if needed.)

7. Do not eat or drink *anything* in the laboratory. This applies to both food and chemicals. The obvious danger is poisoning.

    b. Not so obvious is that you should never touch chemicals. Many chemicals are absorbed through the skin. Wash all chemicals off with large quantities of running water.

    c. Wash your hands thoroughly with soap and water when leaving the laboratory.

8. Avoid breathing fumes of any kind.

    a. To test the smell of a vapor, collect some in a cupped hand. Obtain your instructor's written permission before you smell any chemical. Never smell a chemical reaction while it is occurring.

    b. Work in a hood if there is the possibility that noxious or poisonous vapors may be produced.

9. Never use mouth suction in filling pipets with chemical reagents. *Always* use a suction device.

10. Never work alone in the laboratory. There must be at least one other person present in the same room. In addition, an instructor should be quickly available.

11. Wear shoes in the laboratory. Bare feet are prohibited because of the danger from broken glass. Sandals are prohibited because of the hazard from chemical spills.

12. Confine long hair and loose clothing (such as ties) in the laboratory. They may either catch fire or be chemically contaminated.

   a. A laboratory apron or lab coat provides protection at all times. A lab apron or lab coat is required when you are wearing easily combustible clothing (synthetic and light fabrics).

   b. It is advisable to wear old clothing in the laboratory; it is generally not as loose and flammable as new clothing, and it is not as expensive to replace.

13. Keep your work area neat at all times. Clean up spills and broken glass immediately. Clutter will not only slow your work but lead to accidents. Clean up your work space; wipe all surfaces and put away all chemicals and equipment at the end of the laboratory period.

14. Be careful when heating liquids; add boiling chips to avoid "bumping." Flammable liquids such as ethers, hydrocarbons, alcohols, acetone, and carbon disulfide must never be heated over an open flame.

15. Always pour acids into water when mixing. Otherwise the acid can spatter, often quite violently. Remember that "acid into water is the way that you oughter."

16. Do not force a rubber stopper onto glass tubing or thermometers. Lubricate the tubing and the stopper with glycerol or water. Use paper or cloth toweling to protect your hands. Grasp the glass close to the stopper.

17. Dispose of excess liquid reagents by flushing small quantities down the sink. Consult the instructor about large quantities. Dispose of solids in crocks. *Never return reagents to the dispensing bottle.*

18. Carefully read the experiment and answer the questions in the prelab before coming to the laboratory. An unprepared student is a hazard to everyone in the room.

19. Spatters are common in general chemistry laboratories. Test tubes being heated or containing reacting mixtures should *never* be pointed at anyone. If you observe this practice in a neighbor, speak to him or her or the instructor if needed.

20. If you have a cut on your hand, be sure to cover it with a bandage or wear appropriate laboratory gloves.

21. Finally, and most important, *think* about what you are doing. Plan ahead. Do not cookbook. If you give no thought to what you are doing, you predispose yourself to an accident.

- -

Do you have any diagnosed allergies or other special medical needs (check one)?   Yes_____   No_____ .
If yes, please list them in this space.

I have read and understand the Laboratory Safety Rules and have retained a copy for my reference.

_____   _____
(Name)   (Date)

Laboratory Safety Rules

In the home, the kitchen and bathroom are the sites of most accidents. In a school the chemical laboratory poses similar hazards, and yet it can be no more dangerous than any other classroom if the following safety rules are always observed. Most of them are based on simple common sense.

1. Behave responsibly. The dangers of spilled acids and chemicals and broken glassware created by thoughtless actions are too great to be tolerated.

2. Wear approved eye protection *at all times* in the laboratory and in any area where chemicals are stored or handled. Such protection will protect you against impact and chemical splashes. Goggles are strongly recommended and may be required. The only exception is when explicit instructions to the contrary are given by your instructor.

   a. If you should get a chemical in your eye, first rinse with isotonic sterile solution, then wash with flowing water from a sink or fountain for at least 15 min. Get medical attention immediately.

   b. Do not wear contact lenses in the laboratory, even with safety goggles. Contact lenses prevent rinsing chemical splashes from the eye. Vapors in the laboratory (HCl, for example) dissolve in the liquids covering the eye and concentrate behind the lenses. "Soft" lenses are especially bad as chemicals dissolve in the lenses themselves and are released over several hours.

3. Do not perform any unauthorized experiments. This includes using only the quantities instructed, no more. Consult your instructor if you have any doubts about the instructions in the laboratory manual.

4. Do not smoke in the laboratory at any time. Smoking is not just an obvious fire hazard; it also draws chemicals in laboratory air (both as vapors and as dust) into the lungs.

5. In case of fire or accident, call the instructor at once. Note the location of fire extinguishers and safety showers now so that you can use them if needed.

   a. Wet towels can be used to smother small fires.

   b. In case of a chemical spill on your body or clothing, wash the affected area with large quantities of running water. Remove clothing that has been wet by chemicals to prevent further reaction with the skin.

6. Report all injuries to your instructor at once. Except for very superficial injuries, you will be required to get medical treatment for cuts, burns, or fume inhalation. (Your instructor will arrange for transportation if needed.)

7. Do not eat or drink *anything* in the laboratory. This applies to both food and chemicals. The obvious danger is poisoning.

   a. Not so obvious is that you should never touch chemicals. Many chemicals are absorbed through the skin. Wash all chemicals off with large quantities of running water.

   b. Wash your hands thoroughly with soap and water when leaving the laboratory.

8. Avoid breathing fumes of any kind.

a. To test the smell of a vapor, collect some in a cupped hand.   Obtain your instructor's written permission before you smell any chemical. Never smell a chemical reaction while it is occurring.

b. Work in a hood if there is the possibility that noxious or poisonous vapors may be produced.

9. Never use mouth suction in filling pipets with chemical reagents. *Always* use a suction device.

10. Never work alone in the laboratory. There must be at least one other person present in the same room. In addition, an instructor should be quickly available.

11. Wear shoes in the laboratory. Bare feet are prohibited because of the danger from broken glass. Sandals are prohibited because of the hazard from chemical spills.

12. Confine long hair and loose clothing (such as ties) in the laboratory. They may either catch fire or be chemically contaminated.

a. A laboratory apron or lab coat provides protection at all times. A lab apron or lab coat is required when you are wearing easily combustible clothing (synthetic and light fabrics).

b. It is advisable to wear old clothing in the laboratory; it is generally not as loose and flammable as new clothing, and it is not as expensive to replace.

13. Keep your work area neat at all times. Clean up spills and broken glass immediately. Clutter will not only slow your work but lead to accidents. Clean up your work space; wipe all surfaces and put away all chemicals and equipment at the end of the laboratory period.

14. Be careful when heating liquids; add boiling chips to avoid "bumping." Flammable liquids such as ethers, hydrocarbons, alcohols, acetone, and carbon disulfide must never be heated over an open flame.

15. Always pour acids into water when mixing. Otherwise the acid can spatter, often quite violently. Remember that "acid into water is the way that you oughter."

16. Do not force a rubber stopper onto glass tubing or thermometers. Lubricate the tubing and the stopper with glycerol or water. Use paper or cloth toweling to protect your hands. Grasp the glass close to the stopper.

17. Dispose of excess liquid reagents by flushing small quantities down the sink. Consult the instructor about large quantities. Dispose of solids in crocks. *Never return reagents to the dispensing bottle.*

18. Carefully read the experiment and answer the questions in the prelab before coming to the laboratory. An unprepared student is a hazard to everyone in the room.

19. Spatters are common in general chemistry laboratories. Test tubes being heated or containing reacting mixtures should *never* be pointed at anyone. If you observe this practice in a neighbor, speak to him or her or the instructor if needed.

20. If you have a cut on your hand, be sure to cover it with a bandage or wear appropriate laboratory gloves.

21. Finally, and most important, *think* about what you are doing. Plan ahead. Do not cookbook. If you give no thought to what you are doing, you predispose yourself to an accident.

- -

Do you have any diagnosed allergies or other special medical needs (check one)?   Yes_____   No_____ .
If yes, please list them in this space.

I have read and understand the Laboratory Safety Rules and have retained a copy for my reference.

_____   _____
(Name)   (Date)

COMMONLY USED LABORATORY EQUIPMENT

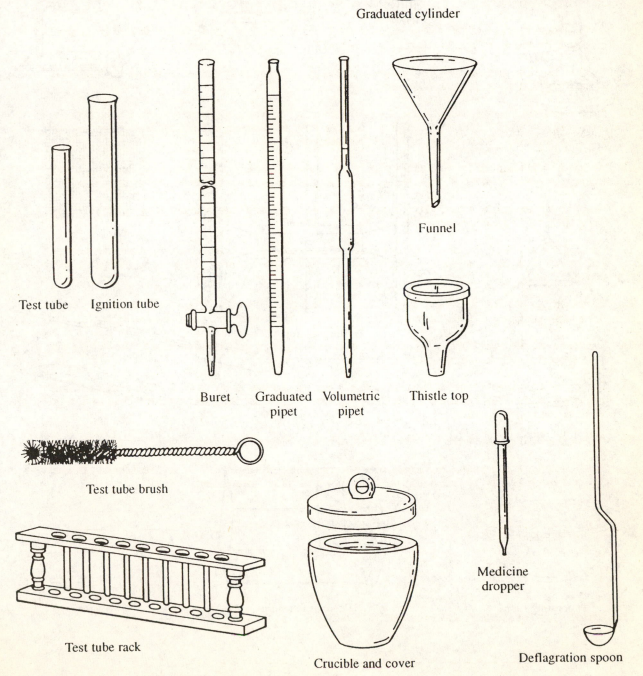

Beaker

Erlenmeyer flask

Florence flask

Graduated cylinder

Wash bottle

Test tube

Ignition tube

Buret

Graduated pipet

Volumetric pipet

Thistle top

Funnel

Test tube brush

Test tube rack

Crucible and cover

Medicine dropper

Deflagration spoon

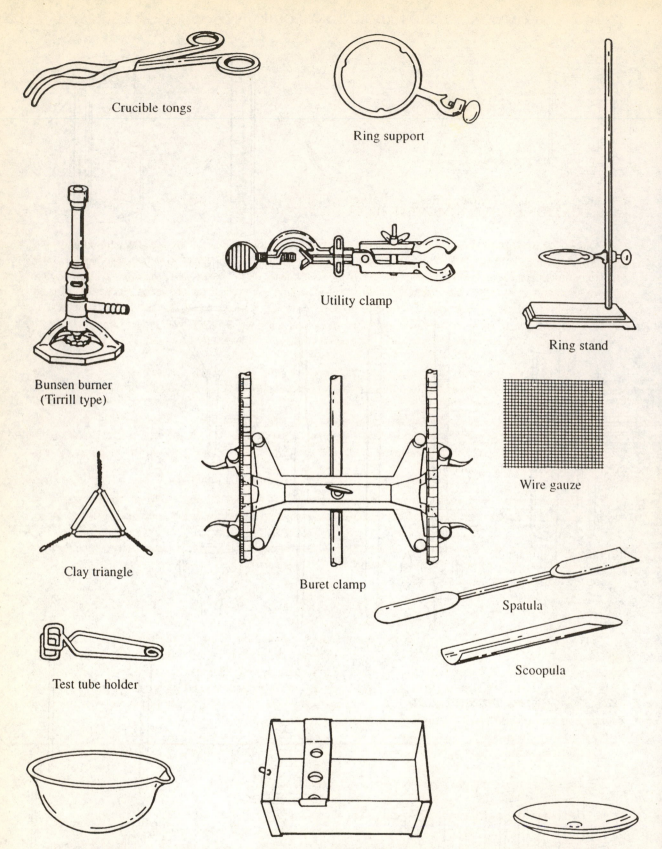

Crucible tongs

Ring support

Ring stand

Bunsen burner
(Tirrill type)

Utility clamp

Clay triangle

Buret clamp

Wire gauze

Test tube holder

Spatula

Scoopula

Evaporating dish

Pneumatic trough

Watch glass

Introduction

AKΨ — AΔ

EVALUATION OF EXPERIMENTAL DATA

One of the most generally accepted axioms in chemistry is that, despite all of the advances in theory during the past 50 years, chemistry is still an experimental science. The vast majority of chemical publications today are experimentally based, with advances being made continually that further our already vast reservoir of information and knowledge. The chemist-scientist of today and the future, therefore, needs to have a thorough grounding in experimental techniques, in how to acquire data, and in how to evaluate the data collected. Only with a knowledge of how to evaluate collected data can any significance be placed on experimental measurements. Precision and accuracy are two essential concepts in the evaluation of data.

In making measurements it is important to recognize and, if possible, estimate sources of error. Experimental error may be classified as either *systematic* (determinate) or *random* (indeterminate).

In systematic errors the cause usually is detectable and can be corrected. This type of error causes measurements to be consistently higher or lower than the actual value. These include errors present in the system itself, errors present in the measuring device, personal error (bias), and gross errors, such as incorrect recording of data or miscalculation.

Random errors are more related to experimental uncertainty than to accuracy. Their sources cannot be identified and are beyond our control. They also tend to fluctuate in a random fashion about a measured value.

One of our main tasks in designing and performing experiments is to reduce or eliminate the effects of systematic error. In this way, only the cumulative effect of the random errors remains, and this may be estimated in terms of precision.

Precision

The key to significance in experimental measurements is repetition. Only with repeated measurements of the density, concentration, or other quantities can the experimenter have some confidence in the significance of the data. Only if a measured quantity can be reproduced repeatedly will the experimenter have that confidence. *Precision* is a quantitative measure of the reproducibility of experimental measurements–how well repeated measurements of the same quantity agree with one another. Precision is frequently measured in terms of the average deviation, which is determined by the following process:

1. Determine the average value from a series of measurements (three or more).

2. Determine the deviation of each measurement from the average value.

3. Determine the average of the deviations without regard to sign.

EXAMPLE: In the determination of the concentration of an unknown acid by titration with standard base, four measurements were made: 0.1025 M, 0.1018 M, 0.1020 M, and 0.1024 M. Compute the average value and the average deviation.

SOLUTION: The average value is computed by summing the four measurements and dividing by four. This yields an average value of 0.1022 M. The individual deviations of each measurement from the average value are 0.0003 (0.1025), 0.0004 (0.1018), 0.0002 (0.1020), and 0.0002 (0.1024). The sum of the four deviations, 0.0011, divided by four yields the average deviation, 0.00028. Since this deviation represents an uncertainty in the measurements, the molarity of the unknown acid is not precisely 0.1022 M, but ranges from 0.1019 M to 0.1025 M and should therefore be reported with the average deviation included–that is, 0.1022 ± 0.0003 M. Further, to make the measurements of precision more useful, the average deviations are put on a percentage basis by determining the relative average deviation. This is the average deviation divided by the average value and multiplied by 100%.

$$\text{Relative average deviation} = \frac{\text{Average deviation}}{\text{Average of measurements}} \times 100\%$$

In the example, the relative average deviation is

$$\frac{0.00028}{0.1022} \times 100\% = 0.27\%$$

A relative average deviation of 0.27% or better (that is, smaller) is a typical expected precision value for an acid-base titration. In general, however, the precision of an experiment varies with the technique and/or apparatus used. A number of variables built into the method or design of the experiment can affect its precision. In a method tested over a long period of time, the results (average value) should not only agree very well with one another (have good precision) but also agree very closely with the true or accepted value (have high accuracy).

Accuracy

The agreement of experimental measurements with the accepted value of a quantity is measured in terms of the error. Error is the difference between the value of a quantity as measured and the accepted value of the same quantity:

$$\text{Error} = \text{Measured value} - \text{Accepted value}$$

When the error in a measurement is put on a relative basis it becomes more useful and is known as the relative error – the error divided by the accepted value and multiplied by 100%. This is usually known as the *percent error*:

$$\% \text{ error} = \frac{\text{Error}}{\text{Accepted value}} \times 100\%$$

In the above example, let us assume that the accepted value of the unknown acid molarity is 0.1014 M as determined by several different experienced experimenters using sound technique. We compute the error and percent error for the determination of the molarity of the unknown acid.

Since the error is the difference between the measured value and accepted value, in this case it is 0.1022 – 0.1014 = + 0.0008. From this error, the percent error is calculated.

$$\% \text{ error} = \frac{+0.0008}{0.1014} \times 100\% = + 0.8\%$$

The only significance of the sign of the error (+ or –) is that the measured value is greater or smaller than the accepted value. In the example above, the percent error, which measures the accuracy of the experiment, is larger than the precision. This is an indication of the existence of systematic error and can be corrected. If all systematic errors have been eliminated, the accuracy should be comparable to the precision of the experiment that measures random error. Thus the accuracy of the experiment is related to the precision (random error) but is also related to possible systematic error.

Propagation of Errors and Deviations

Often the final result of an experiment will be a combination of the measurements of several different items. The errors (or deviations) in all of these measurements are combined to produce the error (or deviation) of the result. If two measurements are added or subtracted, the error of the result is the sum of the absolute errors of the two measurements. If two measurements are multiplied or divided, the error of the result is the sum of the relative errors of the two measurements. Identical rules apply to the propagation of deviations.

EXAMPLE: The density of a solution was determined by first weighing a clean, dry, stoppered flask four times. A mass of 26.413 ± 0.005 g was obtained. 10.00 ± 0.02 mL of the solution was added with a pipet and the stoppered flask was again weighed four times, obtaining a mass of 37.126 ± 0.003 g. Determine the density of the solution and its relative, average, and absolute deviations.

SOLUTION: The mass of the solution is 37.126 g – 26.413 g = 10.713 g. The absolute deviation in this mass is 0.005 g + 0.003 g = 0.008 g and the relative average deviation is (0.008 g/10.713 g) × 100% = 0.07%. The relative average deviation in the volume is (0.02 mL/10.00 mL) × 100% = 0.2%. The density is 10.713 g/10.00 mL = 1.071 g/mL and its relative average deviation is 0.07% + 0.2% = 0.3%. The absolute deviation of the density is (0.3%/100%)(1.071 g/mL) = 0.003 g/mL. Thus, we would report 1.071 ± 0.003 g/mL as the density.

Significant Figures

An understanding of the extent to which the numbers in measured quantities are significant is associated with the evaluation of experimental data. For example, the mass of an object can be measured on two different balances, one a top-loading balance sensitive to the nearest 0.001 g and another a triple-beam balance sensitive to the nearest 0.1 g. These balances have a different uncertainty and precision.

	Top-Loading Balance	Triple-Beam Balance
Quantity	54.236 g	54.2 g
Uncertainty	± .002 g	± 0.1 g
Measured mass	54.236 ± 0.002 g	54.2 ± 0.1 g
Precision	High (2 parts in 54,236)	Low (1 part in 542)

On the top-loading balance with high precision, five significant figures are available. On the triple-beam balance only three figures are significant. Thus significant figures are the numbers about which we have some knowledge.

If no information is available regarding the uncertainty of the measuring device, we may assume that all recorded figures are significant with an uncertainty of about one unit in the last digit. Zeros are significant if they are part of the measured quantity but not if they are used to locate the decimal place. Thus 62.070 has five significant figures while 0.0070 has only two (the first three zeros only locate the decimal place).

In calculations involving measured quantities with different numbers of significant figures, the result must be evaluated carefully with respect to the number of digits retained.

In addition or subtraction, the number of digits retained is based on the least precise quantity. For instance, consider the following summation of masses.

$$
\begin{array}{r}
125.206 \text{ g} \\
20.4 \quad\text{ g} \\
\underline{3.58 \quad\text{ g}} \\
149.186 \text{ g}
\end{array}
$$

Here, the result should be rounded to 149.2 g since the least precise mass is known only to the first decimal place.

In multiplication and division, the number of significant figures retained in a result is equal to the number in the least reliably known factor in the computation. For example, in the determination of the density of an object, the measured mass is 54.723 g while the measured volume is 16.7 mL. The density of the object is 54.723 g/16.7 mL = 3.28 g/mL. Note that only three significant figures can be retained.

When a number is rounded, the last figure retained is increased by one unit if the one dropped is more than five and decreased by one unit if the one dropped is less than five. If the number dropped is a five, the preceding number is rounded up (or left unchanged) in order to make it even. For example,

$$3.276 \rightarrow 3.28; \quad 149.74 \rightarrow 149.7; \quad 5.45 \rightarrow 5.4; \quad \text{and} \quad 5.75 \rightarrow 5.8.$$

THE USE OF THE LABORATORY BALANCE

Today, the use of electronic, top-loading balances is widespread. The instructor will provide you with specific directions on the use of the balance. If you have any questions about using this instrument, be sure to ask your instructor before proceeding further.

Balances are expensive (a $1000 cost is not uncommon), so be careful when using them. Never drop objects on the pan. Never weigh chemicals directly on the pan; use a container such as a weighing bottle, beaker, or weighing paper and weigh by difference. Objects can be weighed directly, but be sure to zero the balance properly beforehand.

When finished, be sure to turn off the balance and check to make sure that no counter weights are left on the register. Always clean up any spillage in the balance area.

THE LABORATORY NOTEBOOK

Your laboratory notebook should contain a permanent record of all your work in the laboratory and your thoughts about each experiment. It should be possible for you to write a complete description of any given experiment at least six months after you have conducted the laboratory work. The following guidelines will help you produce this record. They may seem awkward at first, but with practice they will become second nature and you will be able to use them in future work, both while you are still in college and afterward in your profession.

The laboratory notebook serves at least the following five functions.

1. *It functions as a record of the steps of a procedure.* It is worthwhile to record a brief outline of the experiment that you are going to perform in your notebook. This serves the triple purpose of familiarizing you with the procedure before you perform it, of helping you recognize any portions of the procedure about which you have questions, and of being a reminder of the techniques you used to collect the data. To make this portion of the record of value, you should record the following:

   a. The procedure itself
   b. Any questions that you have about the procedure and their answers
   c. Any procedural changes that occur during the laboratory

Recording the procedure may seem to be a waste of time. Actually, it saves time. Students who attempt to perform experiments without first carefully reading the entire procedure and taking notes frequently discover that they have done the entire experiment incorrectly and must, therefore, repeat it. Fifteen minutes of careful reading before the laboratory period is more efficient than a wasted hour in the laboratory. In addition, you will find that the experiment goes more smoothly and takes less time if you know all the details in advance. Just as you would not attempt to find a building in an unknown city without carefully studying a map and detailed directions, so you should not expect to perform a successful experiment without knowing in advance where you are going. Finally, if you know what you are attempting, your results will be more accurate.

2. *It is a record of all your data.* To make this record easier to keep, you should lay out a data table in your notebook before coming to the laboratory. Allow enough room for several trials, as the initial trials may have to be discarded. This will be especially true with an unfamiliar procedure, particularly if you have not taken careful notes on the procedure before coming to the laboratory.

During the experiment, all data should be recorded in your laboratory notebook *in ink*. Do not feel constrained by the data table you have written. If extra data appear, write them down. Please do not trust your memory. It is heartbreaking to have to repeat an experiment because not enough data were recorded. Under no circumstance should you write data on spare scraps of paper. Take your notebook along—it is after all a *book of notes* on the experiment. If a piece of data is misrecorded, do not obliterate it like this: 3.142. Instead, draw a single line through it: 3.142. You will often find that "bad" data were acceptable after all. Save all calculations for later; record the raw numbers. For example, if you weigh a sample in a beaker and then subtract the mass of the beaker, record the mass of the sample plus beaker and the mass of the beaker. Leave a space in your data table for the mass of the sample. In this way, you will be able to avoid arithmetic mistakes that you might make during the laboratory period. You should also record all your tentative thoughts about the experiment. If the reaction mixture looks like blue jello, say so. If it then changes to violet water you might want to state that you think a chemical reaction occurred.

3. *In many instances, it is a legal document.* If you are involved in research, it can help establish that you did the work first and, hence, should be awarded the patent or granted the right to publish your work. If you are in medicine, it can be used as evidence in your favor in the event of a malpractice suit. If you are in business, it can establish that you incurred the expenses you have claimed on your income tax return.

Remember that an unbound, undated record in pencil is *valueless* for these purposes. Thus you should keep your notes in ink in a bound notebook. Every page should be numbered and dated and every page should be filled. If you wish to start an experiment at the top of a page, you should cross out the blank space on the preceding page.

4. *It is a scratchpad for your calculations.* These calculations should be in a form such that you can interpret them. If your laboratory report is lost or destroyed, you can easily reconstruct it from this information. You also should write out the answers to all the questions in your notebook. In this way, your laboratory report will be a neat and well-organized final draft. Messy reports are usually unacceptable and *always* result in a lower grade.

5. *It is your personal record.* It is not expected to be neat, although it should be orderly enough so that someone else could figure out what you have done. It must be complete—a permanent record of your achievements in the laboratory. The organization is aided by a table of contents in the front.

It is tempting to try to "save time" by shortcutting the above guidelines. This leads to unfortunate consequences such as repeated experiments, unacceptable reports, and a poor understanding of what was achieved in laboratory. In summary, your notebook should be a bound book, kept in ink with all pages numbered and dated. It should contain the following items:

1. A table of contents
2. A brief outline of the procedure
3. Questions about the procedure and their answers
4. Any deviations from the procedure as outlined
5. All the raw data (without calculations) that were obtained
6. Preliminary thoughts about the experiment

7. All calculations performed to determine the final results
8. The answers to all questions asked in the procedure and in the report

THE PRELABORATORY REPORT

Each experiment in this manual includes a prelaboratory report. The prelab report is to be completed before the experiment is begun in the laboratory. Its purpose is to ensure familiarity with the procedure and provide for a more efficient utilization of limited laboratory time. The prelab questions can be answered after a careful reading of the introduction and procedure of the experiment. Sample calculations are sometimes included to provide awareness of data that needs to be collected and how it is treated. Your instructor may prefer to administer prelab quizzes instead of collecting prelab reports.

THE LABORATORY REPORT

A good laboratory report is the essential final step in performing an experiment. It is in this way that you communicate what you have done and what you have discovered. Since it is the only means in many instances of reporting results, it is important that you prepared it properly.

A laboratory report is a final draft. As such it is always written in ink or typed. A typed laboratory report is necessary if your handwriting is hard to read. There must be no erasures or crossed out areas. The initial draft of a laboratory report belongs in your laboratory notebook for two reasons:

1. It is unlikely that you will get everything correct on the first attempt, which means that a first draft written on the report form itself could be very messy.

2. If the report itself is lost or destroyed, you can easily and quickly rewrite the report from the notebook.

It is essential that a laboratory report be neat. Studies have shown that when the same work is submitted in both neat and sloppy form, the neat version makes the better impression. This is true not only in academic work, but also in the business world. Neat work indicates that the writer knows and cares about the subject matter.

All data should be presented with the correct significant figures and units. The omission of units makes it difficult for the reader to know the size of the numbers being reported. And writing down the wrong number of significant figures amounts to lying about the precision of the data. Too many significant figures implies that you know a number more precisely than you actually do.

All questions should be answered with complete and grammatically correct sentences. Abbreviations should not be included in written answers. Read the sentence out loud to make sure that it makes sense.

Your sample computations should be labeled with their purpose–for example, "mass of the liquid." Within the computation, all numbers must have the correct units and the correct number of significant figures.

Laboratory reports that extend to more than one page should either be stapled together or have your name and the page number at the top right of each page. For example: Paul Smith, page 2 of 4 pages. This makes it more difficult for the instructor to inadvertently misplace pages. Paper-clipping or tearing corners to hold pages together is not acceptable. Reports should also be dated.

Graphs

Graphs are used to present the data in picture form so that they can be more readily grasped by the reader. Occasionally a graph is used to follow a trend. Such a case is the graph of pH versus volume of added base in Experiment 19 (page 172). Notice that the best smooth curve is drawn through the data points. This is not the same as connecting the dots; all of the data points will not fall on the line. Often, however, a graph is used to show how well data fit a straight line. The line drawn may either be visually estimated ("eyeballed") or computed mathematically. There are many essential features of a good graph.

1. The axes must be both numbered and labeled. In the sample graph at the end of this section, the ordinate is labeled "Mass of container and liquid (in g)." It is numbered in equal increments from 50 to 120 g. The ordinate is the up-and-down or the vertical axis or the *y*-axis. The abscissa is labeled "Volume of liquid (in mL)." It is numbered in equal increments from 0 to 60 mL. The abscissa is the right-to-left or the horizontal axis or *x*-axis

2. The graph must have a title. The title of the sample graph is "Sample graph for the determination of liquid density." When we speak of graphing, we always mention the quantity plotted on the ordinate first. Thus this is the graph of the mass of container and liquid (in g) versus the volume of the liquid (in mL).

3. The data points are *never* graphed as little dots. You may use instead small circles, small circles with a dot inside, crosses, asterisks, or X's. If dots are used, data are too easily lost on the graph or "created" by stray blobs of ink.

4. Any lines that appear on the graph in addition to data points should be explained. Thus the line drawn is explained in the title as "visually estimated best straight line."

5. The scales of the axes should be adjusted so that the graph fills the page as much as possible. Thus this graph starts at 50 g since there are no data of less mass and it would be foolish to have that much wasted paper. It would also be harder to read the graph if the scales always had to start with zero.

SAMPLE LABORATORY EXPERIMENT REPORT

The various stages of a hypothetical laboratory experiment are shown on the pages that follow.

1. Lab manual instructions
2. Notebook entries of the data
3. Final lab report
4. Required graph

Experiment 0: Liquid Density Determination–Finagle's Method

The object of this experiment is to determine the density of a liquid using graphical techniques. We shall carefully weigh known volumes of the liquid in a preweighed dry flask and then plot these masses against the volume of liquid weighed. The slope of the graph will be the liquid's density.

Procedure

Carefully clean and dry a 100 mL Erlenmeyer flask and a rubber stopper to fit it. Weigh the flask to the nearest 0.1 g. Using a 10 mL pipet, transfer 10.00 mL of the unknown liquid into the flask. With the stopper in place, weigh the flask and liquid to the nearest g. Repeat this procedure until you have added a total of 50.00 mL. Now, using a 25.00 mL pipet, remove 25.00 mL of liquid. Reweigh the stoppered flask. Construct a graph of mass of flask and liquid vs. volume of liquid and determine the density of the liquid from the slope of the line.

Notebook Page

LIQUID DENSITY DETERMINATION

9/30/96

	TRIAL 1	TRIAL 2	TRIAL 3	AVERAGE
MASS OF FLASK, g	52.2	51.5	51.1	51.6
FLASK + 10.00 mL	65.3	64.2	66.0	65.2
+ 20.00 mL	78.5	79.0	78.1	78.5
+ 30.00 mL	90.5	91.0	90.1	90.5
+ 40.00 mL	106.7	106.3	107.2	106.7
+ 50.00 mL	118.0	118.2	117.6	117.9
+ 25.00 mL	87.5	87.2	87.9	87.5

SLOPE = $(120.0 - 52.0)/(51.1 - 0.2) = 68.0/50.9$

      = 1.34 g/mL

CHECK $(118.0 - 52.2)/50.0 = 65.8/50.0$

                  = 1.32 g/mL

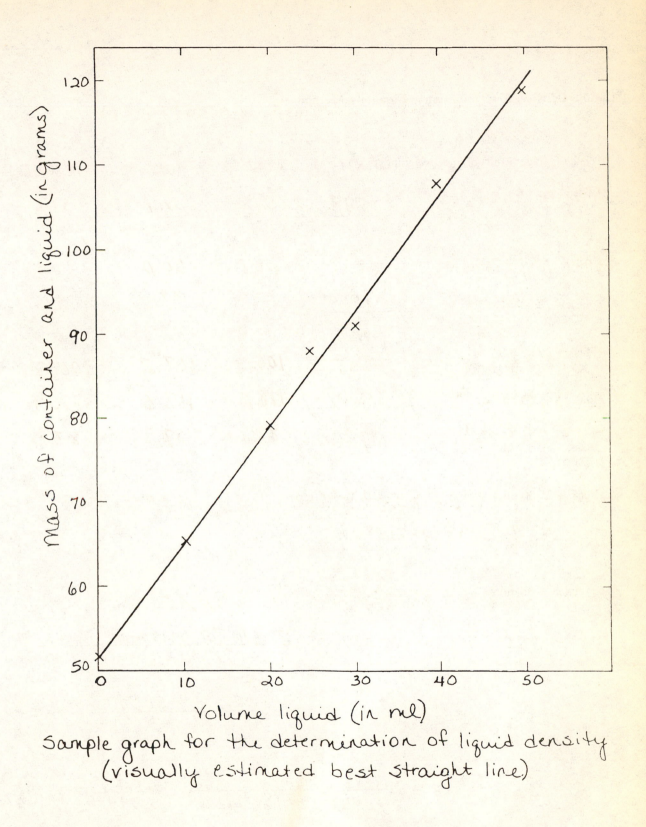

Sample graph for the determination of liquid density
(visually estimated best straight line)

Report   Name _____ Section _____

Liquid Density Determination - Finagle's Method

Data (average results of three weighings)

Mass of flask   *51.6 g*

Flask + 10.00 mL   *65.2 g*

Flask + 20.00 mL   *78.5 g*

Flask + 30.00 mL   *90.5 g*

Flask + 40.00 mL   *106.7 g*

Flask + 50.00 mL   *117.9 g*

Flask + 25.00 mL   *87.5 g*

Density of liquid   *1.34 g/mL*

Sample Calculations

$$\text{AVERAGE MASS OF FLASK} = \frac{\text{SUM OF MASS OF THREE TRIALS}}{\text{NUMBER OF TRIALS}}$$

$$= \frac{52.2\,g + 51.5\,g + 51.1\,g}{3} = 51.6\,g$$

$$\text{SLOPE OF LINE} = \frac{\text{RISE}}{\text{RUN}} = \frac{120.0\,g - 52.0\,g}{51.1\,mL - 0.2\,mL} = \frac{68.0\,g}{50.9\,mL} = 1.34\,g/mL$$

A GRAPH OF THE DATA IS ATTACHED.

1 Measurements and Density

INTRODUCTION

As we have emphasized in the Preface, chemistry is very much an experimental science in which careful and accurate measurements are the very essence of meaningful experimentation. It is therefore essential for the beginning student of chemistry to learn how scientific measurements are carried out properly through the use of common measuring instruments. It is equally important for the student to acquire an appreciation of the significance of measurements and to apply learned technique to a common specific experiment.

In the following experiment you will become familiar with how mass and volume measurements are carried out and how an evaluation of the measurements is reflected in the number of significant figures recorded. These mass and volume measurements will then be used to determine the density of (1) a metal bar and (2) a salt solution by two different methods. Finally, the results of the density measurements will be evaluated with respect to their precision and accuracy.

The density of an object is one of its most fundamental and useful characteristics. As an intensive property it is independent of the quantity of material measured since it is the ratio of the mass of an object to its volume. The density of an object can be determined by a variety of methods. In this experiment you will practice using a balance to measure mass. In addition, you will learn how to measure volume using a graduated cylinder and a pipet and learn how to calibrate the pipet. A comparison of the results allows for the calculation of the relative average deviation, which is a measure of the precision of the experiment. Also, in the case of the metal bar, the results of measuring the density of the bar may be compared with the accepted density value for the bar. Thereby the relative error (a measure of accuracy) for the density of the bar may be determined. The sections in the Introduction to this laboratory manual pertaining to precision, accuracy, significant figures, and the laboratory notebook should be studied carefully before performing this experiment.

SAFETY PRECAUTIONS

Review the safety rules on pages 1 and 2.

Take special care in inserting the bar into the graduated cylinder. **Do not drop it in!** The glass cylinder may break.

Pipeting should always be done using a suction device. **Never suction by mouth.**

PROCEDURE

Record all measurements in your laboratory notebook in ink. The proper use of a sensitive balance is critical to useful mass measurements. Also, pipeting is a very useful, accurate, and common method for transferring exact volumes of liquids. Therefore, the instructor should demonstrate good balance and pipet techniques to the class at the beginning of the laboratory period. Please note that when a portion of the experiment contains the instruction "Repeat . . . twice," each portion is to be performed all the way through three times: initially and two repetitions.

PART I: MEASUREMENTS

A. Mass Measurements

After balance instruction, you will be assigned or allowed to select a balance for use during the experiment.

1. Zero the balance after cleaning the pan.
2. Measure the mass of a clean dry 50 mL beaker to the nearest ±0.001 g.
3. Record, in ink, your observation **directly** into the lab notebook.
4. Remove the beaker from the pan. Again, clean the balance pan and zero the balance.
5. Weigh the same beaker as before (step 2) and record the result.
6. Repeat steps 4 and 5 one more time.
7. From the three mass measurements, calculate the average mass of the beaker.
8. Repeat steps 4 and 5 using a second balance (just one weighing).
9. Repeat steps 4 and 5 using a third balance (just one weighing).

B. Volume Measurements

Use of a pipet: In order to accurately measure a liquid volume using a pipet, you must consider several things. Most volumetric pipets are designed to deliver rather than to contain the specified volume. Thus a small amount of liquid remains in the tip of the pipet after transfer of liquid. This kind of pipet is marked with the letters "TD" somewhere on the barrel above the calibration line. Also, for purpose of safety, **never** pipet by mouth; that is, never use your mouth to draw liquid into the pipet. **Always** use a suction device.

Use a clean but not necessarily dry 20 or 25 mL pipet. Rinse the pipet several times with small portions of the liquid to be transferred. To measure the desired volume, a volume of liquid greater than that to be measured is needed in order to keep the pipet tip under the liquid surface while filling.

While holding the pipet vertically, squeeze the air out of the suction device and hold it against the large end of the pipet, tight enough to obtain a seal. Keep the suction device evacuated and dip the pipet tip below the surface of the liquid, but do not touch the bottom of the container. (A chipped tip causes error.) Now release the suction device gently and allow liquid to fill the pipet until it is one to two cm above the calibration line etched onto the upper barrel. Quickly remove the suction device and cover the end with your index finger before the liquid level falls below the line (some practice may be necessary). Wipe the outside of the tip with a clean piece of towel or tissue. With the tip touching the wall of the source container above the liquid level, allow it to drain until the meniscus rests exactly on the line. Now hold the pipet over the sample container and allow it to drain, but be careful to avoid loss from splashing. When the swollen part of the pipet is nearly empty, touch the tip to the wall of the container and continue draining. When the liquid level falls to the tip area, hold the tip to the glass for an additional 20 seconds and then remove. Do **not** blow out the remaining liquid.

1. Measure the temperature in the laboratory. Your instructor will provide you with the density of water at this temperature.
2. Use the same 50 mL beaker from Section A for determining the mass of each aliquot of water. Rather than reweighing the empty beaker, the average mass of the beaker determined in Section A may be used as the mass of the dry beaker.
3. Measure 20 or 25 mL of water (depending on the size of pipet available) into the 50 mL beaker.
4. Record the volume of water measured with the pipet to the appropriate number of significant figures.
5. Record the number of significant figures in the volume measurement.
6. Weigh the beaker and water to the nearest mg (±0.001 g).
7. Calculate the mass of water in the beaker.
8. Use the mass and density of water to determine the volume of water measured.
9. Repeat steps 3–8 using a 50 or 100 mL graduated cylinder instead of the pipet to measure the 20 or 25 mL of water. Repeat steps 3–8 again using a graduated 50 mL beaker to measure the water.

C. Calibration of a Volumetric Pipet

Your instructor may suggest that you calibrate your pipet. When you do so, use a pure liquid of known temperature. Usually water is most convenient to use. Pipet distilled water into a preweighed flask with stopper. After pipeting, weigh the flask, stopper, and add water. Use the measured mass of water and its density value at that temperature, given by your instructor or from a handbook, to calculate the volume of the liquid delivered by the pipet to four significant figures. Do at least three separate determinations.

PART II: DENSITY

A. Density of a Metal Bar (Use the same metal bar for all trials.)

1. Zero your balance. Weigh a metal bar on a balance sensitive to the nearest mg (±0.001 g). Repeat the entire weighing operation twice. Do not allow the first measurement that you obtain to influence subsequent measurements that you make. Make sure you zero the balance before proceeding with each measurement.

2. Determine the volume of the metal bar by each of the following methods, making at least three measurements for each method. Do not allow the first measurement to influence subsequent measurements as your data will then be less significant for the purpose of measuring the precision of this experiment.

Method I

Insert the bar into a graduated cylinder filled with enough water so that the bar is immersed. Note and record as precisely as possible the initial water level, and the water level after the bar is immersed. Read the lowest point of the meniscus in determining the water level and estimate the volume to one digit beyond the

2 Formula and Composition of a Hydrate

INTRODUCTION

Hydrates are crystalline compounds in which one or more molecules of water are combined with each formula unit of a salt—the anhydrous compound. This water of hydration often is not bound tightly into the crystalline structure of the anhydrous compound and can usually be driven off by heating a sample of the hydrate with a Bunsen burner for 5 to10 min. If the hydrate is colored, a color change usually results upon heating as the anhydrous salt forms. For example,

$$CuSO_4 \cdot 5H_2O(s) \longrightarrow CuSO_4(s) + 5\ H_2O(g)$$
$\quad$ *blue* $\qquad\qquad\qquad$ *white*

$$CoCl_2 \cdot 6H_2O(s) \longrightarrow CoCl_2(s) + 6\ H_2O(g)$$
$\quad$ *red* $\qquad\qquad\qquad$ *blue*

$$NiSO_4 \cdot 7H_2O(s) \longrightarrow NiSO_4(s) + 7\ H_2O(g)$$
$\quad$ *green* $\qquad\qquad\qquad$ *yellow*

In some cases, the hydrate can form spontaneously from the anhydrous salt if sufficient moisture is present in the air. In yet other cases, the hydrate will lose its water of hydration spontaneously at room temperature. In this experiment the hydrates to be investigated will be limited to those that are stable at room temperature but that decompose to the anhydrous form on heating. Please note that many hydrates—and their anhydrous forms—are white crystalline salts, so that a color change may not occur. After heating, the anhydrous salt must be cooled in the absence of moisture so that the hydrate does not re-form before weighing. A desicooler may be used for this purpose.

From the masses of hydrate, anhydrous salt, and eliminated water, the formula of the hydrate may be calculated if the formula of the anhydrous form or anhydrate is known. If the identity of the anhydrous compound is unknown, the percentage of water in the hydrate may be calculated instead. Proper technique in handling equipment is essential in this analysis, and careful heating and cooling of the sample are necessary if precise and accurate results are to be obtained. The purpose of Part A of the experiment is to give you an opportunity to check the quality of your technique as it is reflected by the precision and accuracy of your results. With good technique, the percentage of water in an unknown hydrate may be determined with an accuracy of 1% or better. All masses must therefore be determined as carefully as possible and recorded to the nearest 0.001 g.

SAFETY PRECAUTIONS

Review the safety rules on pages 1 and 2.

It is not usually obvious from an object's appearance that it is hot. Severe burns can result if you are not very careful when handling a Bunsen burner, tripod, clay triangle, or crucible when these are hot.

Discard the anhydrous form after each trial in the manner and location recommended by your instructor. Many hydrates dissolve in their water of hydration on initial heating, forming a very concentrated solution which spatters easily and sometimes violently. Therefore, heat gently initially and strongly only after most of the water has been driven off.

PROCEDURE

A. Verification of the Formula of a Known Hydrate: $CaSO_4 \cdot 2H_2O$

1. The Bunsen burner should be adjusted so that the mixture of air and gas produces a flame with a blue (oxidizing) cone about 2 to 3 cm above the top of the burner. Proper burner adjustment is crucial to good results in this experiment–so take special care with it. Consult with your instructor if you have difficulty in adjusting your burner.

2. Clean a crucible and lid thoroughly, place them on a clay triangle suspended on a tripod, and heat them for about 5 min. (The crucible should be about 2.0 cm above the oxidizing cone of the burner flame.) Then use tongs to remove the crucible and the lid and place them in your desicooler to cool to room temperature. When the desicooler no longer feels warm, remove the crucible and lid with the tongs and weigh them. Under no circumstances should you touch the crucible with your fingers at any time during the cooling and weighing steps. Please note also that you *must* wait for the crucible to cool to room temperature before weighing it to avoid error due to buoyancy (warmer objects weigh less than cooler ones). In order to use your time efficiently, clean and heat a crucible for your second sample, while the first one is cooling.

3. When the crucible mass is obtained (to the nearest 0.001 g), add 0.7 to 1.7 g of $CaSO_4 \cdot 2H_2O$ to the crucible and weigh again. Then heat the crucible strongly for 4 to 8 min (the crucible should be about 2.0 cm above the oxidizing cone of the burner flame). Parts of the crucible should be dull orange during at least the last 3 min it is being heated. The lid of the crucible should be *in place* during the heating. Be sure to heat the top and sides of the crucible to bake the water out of any solid that may have spattered there. Cool and weigh the crucible. Do not dispose of the sample until after calculating the formula of the hydrate.

4. Calculate the formula of the hydrate as a ratio: moles H_2O/moles $CaSO_4$.

5. Repeat steps 2 and 3 with a second sample.

6. If the formulas of your two samples do not agree within 2% or if either one differs from the correct value by more than 4%, consult the instructor before you continue. You may be asked to run another trial, since the procedure may not have been followed flawlessly.

DISPOSAL: **Calcium sulfate:** Put in a waste bottle labeled for this salt.

B. The Composition of an Unknown Hydrate

1. Clean your crucibles after completing Part A and heat them carefully as before. Add 0.5 to 0.6 g of unknown to your crucible and heat as before. Do this very gently at first to minimize spattering. Cool and weigh the crucible.

2. Repeat with a second sample.

3. Calculate the percent by mass of water in the unknown hydrate.

4. Repeat with additional samples until two successive calculations of the percent water in the hydrate have a relative average deviation of no more than 3%.

DISPOSAL: **Unknown solids:** Put into a labeled (Experiment 2) waste container.

Report   Name _____ Section _____

A. Formula of CaSO₄ Hydrate

	Trial 1	Trial 2
Mass of crucible, lid + hydrate, g	_____	_____
Mass of crucible and lid, g	_____	_____
Mass of hydrate, g	_____	_____
Mass of crucible, lid + anhydrous CaSO₄, g	_____	_____
Mass of anhydrous CaSO₄, g	_____	_____
Moles of anhydrous CaSO₄	_____	_____
Mass of water eliminated, g	_____	_____
Moles of water eliminated	_____	_____
Formula of hydrate (= moles H₂O/moles CaSO₄)	_____	_____
Average formula	_____	

Sample Calculations

Report   Name _____ Section _____

B. Composition of an Unknown Hydrate

Number of unknown _____

_____

	Trial 1	Trial 2	Trial 3
Mass of crucible, lid + hydrate, g	_____ g	_____ g	_____ g
Mass of crucible and lid, g	_____ g	_____ g	_____ g
Mass of hydrate, g	_____ g	_____ g	_____ g
Mass of crucible, lid + anhydrous compound, g	_____ g	_____ g	_____ g
Mass of anhydrous compound, g	_____ g	_____ g	_____ g
Mass of water eliminated, g	_____ g	_____ g	_____ g
Percent by mass of water in unknown	_____ %	_____ %	_____ %

Average percent by mass of water   _____ %

Sample Calculations   (Place these on the previous page.)

QUESTIONS

1. In procedure B, if the empty crucible is not heated long enough to completely dry it prior to weighing, what will be the effect on the calculated percent water in the unknown (larger or smaller)? Explain.

2. In procedure B, if the heated crucible containing the unknown is allowed to cool on the lab bench instead of in the desicooler, what will be the effect on the calculated percent water in the unknown (larger or smaller)? Explain.

3. In procedure B, if the crucible containing the unknown is not heated long enough, what will be the effect on the calculated percent water in the unknown (larger or smaller)? Explain.

Prelab   Name _____ Section _____

Formula and Composition of a Hydrate

    1.   A student heats a crucible and sample of a hydrate as directed in the procedure, then allows the sample to cool in the crucible while sitting out on the lab bench rather than in a desicooler. What will happen to the mass of the contents of the crucible as it cools? Explain why this happens.

    2.   What is the safety precaution associated with handling the crucible, clay triangle, and tripod?

    3.   At the beginning of the procedure, why must the crucible be cleaned before weighing? Why must the crucible be heated and allowed to cool before weighing?

    4.   A 1.084 g sample of hydrate is heated, leaving 0.762 g of anhydrous residue.
        a. How many grams of water were driven off?
        b. How many moles of water were driven off?
        c. What is the percent mass of water in the hydrate?

    5.   In the first trial of part A of the procedure, the ratio of moles H_2O/moles $CaSO_4$ is found to be 1.93. In the second trial the ratio was found to be 1.98. Calculate the relative average deviation of the two trials. Should the student run another trial?

3 Physical Changes and Chemical Reactions

INTRODUCTION

The purpose of this experiment is to observe carefully several changes in matter and to determine if they are physical changes or chemical reactions.

In a *physical change* the appearance of a substance changes, but its composition and identity are unaltered. Examples of physical changes include the boiling of water to produce steam, the filing of metal to produce powder or filings, and the dissolving of sugar in water to form sugar syrup. In most cases, one or two simple processes are all that are needed to reverse the physical change. For example, sugar may be isolated from sugar syrup merely by evaporating the water. But it is the unaltered nature of the products that indicates a physical change, not the ease of reversing the process. For example, sharpening a pencil is a physical change—you start and end with graphite and wood—even though reconstructing the pencil from the shavings is a very difficult process. *Physical properties* are such things as color, state (solid, liquid, or gas), density, and melting and boiling points. They do *not* involve a chemical reaction.

In a *chemical reaction* a change in the composition and identity of a substance occurs. Some chemical reactions are the burning of wood to form carbon dioxide and water, the rusting of iron to form iron oxide, and the heating of limestone to form lime and carbon dioxide. The substances present before a chemical reaction occurs are called *reactants*. The substance or substances formed are called the product(s) and these product(s) have physical and chemical properties that are different from those of the reactants. *Chemical properties* are displayed when a substance undergoes a chemical reaction to produce products. Reversing a chemical reaction usually requires an involved process of several steps. For example, green plants use photosynthesis—a complex series of reactions that science has not yet duplicated—to make wood from carbon dioxide and water.

A chemical reaction is indicated by any of the following observations:

1. Change of color
2. Production of heat, light, or sound
3. Evolution of a gas
4. Formation of a solid precipitate from a liquid or solution

These observations are not infallible indications that a chemical reaction has occurred. For example, when ice forms from liquid water, it may seem that a solid precipitate forms from a liquid, but a physical change rather than a chemical reaction has occurred. In this case, we would classify the change as physical based on the fact that both ice and water are the same substance—the compound of formula H_2O—in either the solid or liquid form.

SAFETY PRECAUTIONS

Review the safety rules on pages 1 and 2.

Wear goggles at all times.

Never taste any chemical.

Smell chemicals by fanning some of the vapors toward your nose and then only when specifically directed to do so. Never smell a reaction mixture when it is being heated or when it is reacting. Obtain explicit permission from your instructor in writing before smelling any chemical.

Some chemicals are poisonous! If you must touch chemicals or if you inadvertently get chemicals on you, wash immediately with soap and water. Report this contact to your instructor.

Use small quantities of chemicals as directed. Compare the amounts you take with those in labeled test tubes prepared by your instructor or weigh the chemicals. Not only are large quantities of chemicals hazardous, but the reactions that occur are difficult to observe with precision.

Stay at a distance from a reaction mixture while it is being heated.

Never use anything but a glass stirring rod to mix chemicals. If you can, mix chemicals by gently tapping the side of a test tube with your finger. (Your instructor will demonstrate this technique.)

Dispense reagents either with the dropper or spatula supplied with each bottle or by pouring out a small quantity of liquid or solid reagent. Take only what you need in order to minimize waste. **Do not** return excess reagent to a bottle; it may contaminate the rest of the contents. Instead, discard it in the place and manner designated by your instructor.

PROCEDURE

Observe what changes occur in each of the following procedures and classify the change as a physical change or a chemical change. The process of observation involves the following.

1. Look carefully at the reactants, the products, and the change.
2. Listen to the change as it occurs.
3. Cautiously smell (see the Safety Precautions) the reactants and the products, if instructed to do so.
4. Observe subsequent changes of the products in other parts of the experiment.

A. Changes Caused by Heat

1. Heat a very small quantity of each of the following substances in a Bunsen burner flame. For granular substances, place an amount equal in volume to 1 grain of rice on a deflagrating spoon and place this in the flame. For the metals, hold a short (0.5 cm or less) strip in the flame with tongs. Some of the metals burn very brightly; **protect your eyes. Do not smell any of the gases produced in this part. Work in the hood.**

Granular	Metal
Sugar	Copper wire (save the product for Part B)
Iodine	Steel wool
Sulfur	Magnesium
Sand	Zinc
Ammonium chloride	Nichrome wire (save the product for Part B)

DISPOSAL of residue:
   Sucrose: Put any solid into trash container and flush any solution down sink.
   Iodine, sand, zinc: *Recycle* in a bottle labeled for that substance.
   Sulfur: *Dispose* in a waste bottle labeled for *reducing agents—solids*.
   Magnesium (MgO), steel wool, ammonium chloride: None are considered to be hazardous and may be disposed in the classroom trash container.

2. **Work in the hood during this part of the experiment.**

   a. Place 0.2 g or less of each of the following solids, one at a time, in a medium Pyrex test tube fitted with a rubber stopper and a length of glass tubing. **Do not use a large quantity of solid. Compare the quantity you take with that displayed in a labeled test tube.** Clamp the test tube on a ring stand and heat (gently at first and then quite strongly) with a handheld Bunsen burner as shown in Figure 3-1. Observe the solid before and after heating.

   b. Wet a pieces of red and blue litmus paper with distilled water. Hold the two pieces together in the gas stream with a pair of tongs. Note any color changes in the litmus paper. Acids turn blue litmus paper red. Bases turn red litmus paper blue.

   c. Collect some of the gas evolved by holding a small test tube with a test tube holder over the open end of the glass tubing. Remove the Bunsen burner immediately when you see gas escaping into the air from the mouth of the collecting test tube. Stopper the collecting test tube quickly to prevent the escape of gas and observe the color of the gas. **Do not smell these gases**—some are poisonous!

   d. Introduce a glowing splint into the collecting test tube (which is still being held by a test tube holder) and observe. Are the gases different? Place a glowing splint in a test tube filled with air for comparison. If the splint glows more brightly or ignites when inserted in the tube, oxygen is present. If there is an audible pop, then hydrogen is probably present. If the splint is extinguished, a nonflammable gas such as CO_2, N_2 or Cl_2 is present.

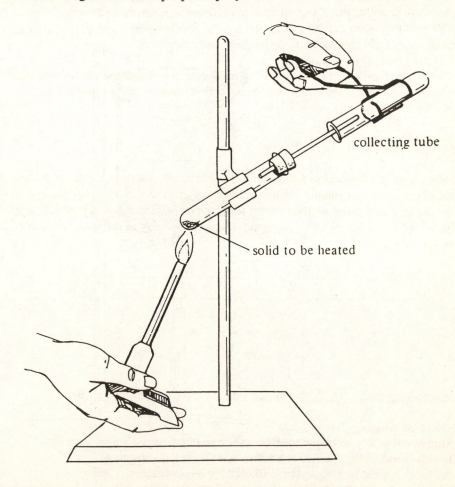

collecting tube

solid to be heated

Figure 3-1. Apparatus for Heating a Solid

e.   Observe the following solids:

Potassium chlorate, $KClO_3$

$KClO_3$ with a very small quantity of manganese dioxide, MnO_2. Mix the two solids by gently tapping the test tube with your finger. **Do not stir the two solids.**

Copper(II) carbonate, $CuCO_3$

Iron(III) nitrate nonahydrate, $Fe(NO_3)_3 \cdot 9H_2O$

Copper(II) sulfate pentahydrate, $CuSO_4 \cdot 5H_2O$

Save the copper(II) sulfate pentahydrate product and the iron(III) nitrate nonahydrate product in separate test tubes for Part B.

DISPOSAL

Unused potassium chlorate: *Dispose* in a waste bottle labeled *oxidizing agents—solids*.
Unused manganese dioxide, copper(II) carbonate: *Dispose* in a waste bottle labeled *heavy metals—solids*.
Iodine: *Recycle* or *dispose* in a bottle labeled for iodine.
Potassium chlorate-manganese dioxide residue: *Dispose* in a waste bottle labeled *heavy metals—solids*.

3.   Support a 150 mL beaker with wire gauze on a ring stand. Boil 100 mL of distilled water in the beaker. Remove the Bunsen burner. Place 0.5 g of iodine in an evaporating dish and cover with a watch glass, convex side down. Place a small amount of ice in the watch dish. The apparatus is shown in Figure 3-2. Observe the process and the fine crystals on the bottom of the watch glass and in the evaporating dish.

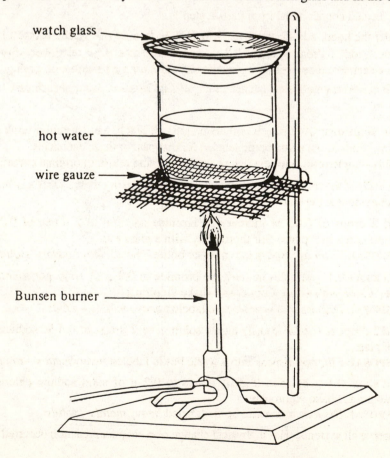

watch glass

hot water

wire gauze

Bunsen burner

Figure 3-2. Apparatus for Heating Iodine

B. Changes Caused by Pure Liquids and Solutions

1. Observe carefully what happens in each of the following cases. Try especially to explain what happens when you add water to the products of heated solids. For instance, was water the only substance that was driven off by heating?

   a. Mix about 0.1 g of $CuSO_4 \cdot 5H_2O$ with 10 drops of water in a small test tube. Stir well.

   b. Mix about 0.1 g of the copper(II) sulfate pentahydrate product from Part A, step 2, with 10 drops of water in a small test tube. Stir well.

   c. Mix about 0.1 g $Fe(NO_3)_3 \cdot 9H_2O$ with 10 drops of water in a small test tube. Stir well.

   d. Mix about 0.1 g of the iron(III) nitrate nonahydrate product from Part A, step 2, with 10 drops of water in a small test tube. Stir well.

   e. Observe. Has a physical change or a chemical reaction occurred in each case?

   f. Add 5 drops of 6 M NH_3(aq) to each test tube and stir well. Observe. Does a physical change or a chemical reaction occur in each case?

   DISPOSAL
   Solutions: *Dispose* in a waste bottle labeled *heavy metals—solutions.*

2. a. In each of five small test tubes, place about 0.1 g of

      copper metal in two of the test tubes

      zinc metal

      nichrome wire (reuse the piece from Part A, step 1)

      heated copper metal from Part A, step 1

   b. Under the hood, add 10 drops of 6 M HNO_3 to only one of the two test tubes containing copper metal. Add 10 drops of 6 M HCl to each of the other four test tubes. Stir each mixture well and allow each test tube to stand in the test tube rack for 5 min before observing.

   c. Observe. Does a physical change or a chemical reaction occur in each case?

   DISPOSAL
   Acid solutions: Add phenolphthalein and 1 M NaOH until solution turns pink (a precipitate may form). Pour into a waste bottle labeled for disposal of these chemicals.
   Solids—nichrome, copper: Rinse with water and return to original container.

3. Observe each of the following carefully while you are mixing them. Does anything occur during mixing that is not evident at the end?

   a. Add 2 drops of 0.01 M mercury(II) bromide solution to 2 drops of 0.02 M potassium iodide solution on a spot plate. Stir thoroughly with a glass rod.
      DISPOSAL *Filter* or *dispose* into a waste bottle labeled *heavy metals—solutions.*

   b. Add about 0.1 g of solid mercury(II) bromide to 0.2 g of solid potassium iodide in a small test tube. Close with a cork stopper and shake vigorously.
      DISPOSAL *Dispose* in a waste bottle labeled *heavy metals—solids.*

   c. Add 2 drops of 0.2 M lead(II) nitrate solution to 2 drops of 0.4 M sodium chloride solution on a spot plate.
      DISPOSAL *Filter* or *dispose* into a waste bottle labeled *heavy metals—solutions.*

   d. Add about 0.1 g of solid lead(II) nitrate to 0.2 g of solid sodium chloride. Close with a cork stopper and shake vigorously.
      DISPOSAL *Dispose* in a waste bottle labeled *heavy metals—solids.*

   e. Observe all systems. Has a physical change or a chemical reaction occurred in each case?

Report Name _____ Section _____

Summarize your recorded observations in the following table. Use one or two word descriptions and abbreviations in each blank space–for example, *gr* = green, *or* = orange, *rd* = red, *y* = yellow, *vi* = violet, *wh* = white, *cls* = colorless, *bu* = blue, *bk* = black, *out* = puts out splint, *flare* = splint flares up, *B→R* for blue to red, *R→B* for red to blue, *ppt* for precipitate, and *L→S* for a liquid to solid change of state. In the column labeled "brief explanation," list those items that allow you to determine whether the process is a physical change or a chemical reaction.

A. Changes Caused by Heat

	Heat Light Sound	Color Change	Change of State	Gas Color	Splint Test	Litmus Test	Ppt. Color	Addn'l Observ.	Chem. or Phys.	Brief Explan-ation
Sugar										
Iodine										
Sulfur										
Sand										
NH_4Cl										
Cu wire										
Steel wool										
Magnesium										
Zinc										
Nichrome wire										
$KClO_3$										
$KClO_3 + MnO_2$										
$CuCO_3$										
$Fe(NO_3)_3 \cdot 9H_2O$										
$CuSO_4 \cdot 5H_2O$										
I_2										

Report _____   Name _____ Section _____

B. Changes Caused by Pure Liquids or Solutions

	Heat Light Sound	Color Change	Change of State	Gas Color	Splint Test	Litmus Test	Ppt. Color	Addn'l Observ	Chem. or Phys.	Brief Explan- ation
$CuSO_4 \cdot 5H_2O$										
$CuSO_4 \cdot 5H_2O$ product										
$Fe(NO_3)_3 \cdot 9H_2O$										
$Fe(NO_3)_3 \cdot 9H_2O$ prod.										
$Cu + HNO_3(aq)$										
$Cu + HCl(aq)$										
$Zn + HCl(aq)$										
Nichrome $+ HCl(aq)$										
Heated $Cu + HCl(aq)$										
$HgBr_2(aq) + KI(aq)$										
$HgBr_2(aq) + KI(s)$										
$Pb(NO_3)_2 + NaCl(aq)$										
$Pb(NO_3)_2 + NaCl(s)$										

Report   Name _____ Section _____

QUESTIONS (Support your answers with your observations.)

1. In Part A, step 1, were the changes caused by heat physical or chemical?

2. In Part A, Step 2, are the gases produced the same? What gas do you believe you collected from each solid? What evidence supports your answer?

3. In Part B, step 2, which of the metals used reacted in the presence of HCl?

4. In Part A, step 3, was the change a physical or chemical change? What observations lead you to this conclusion?

5. What do you believe are the identities of the solids produced by heating $CuSO_4 \cdot 5H_2O$ and $Fe(NO_3)_3 \cdot 9H_2O$? What experimental evidence supports your answer?

Prelab   Name _____ Section _____

Physical Changes and Chemical Reactions

1. How can a physical change be distinguished from a chemical change?

2. What observations would indicate that a chemical change has occurred?

3. What gas was probably produced by a chemical reaction if a glowing splint ignites when placed in the test tube?

4. How does litmus paper identify the presence of an acid or base?

5. Describe the safety precautions associated with smelling a chemical.

4 Types of Chemical Reactions

INTRODUCTION

A large number of inorganic chemical reactions may be classified into one of four general categories: (1) combination, (2) decomposition, (36) single displacement, and (4) double displacement (metathesis). The purpose of this experiment is to examine representative reactions in each category that can be tested readily in the laboratory. In this way, an empirical correlation of theoretical principles with experimental observations will be realized.

Combination Reactions

Combination reactions consist of the direct union of two simple substances to form a third that is more complex. This process can be represented generally by $A + B \longrightarrow AB$. Examples include the following:

1. The combination of a metal with a nonmetal to form a compound:

$$Fe + S \longrightarrow \underset{\text{iron(II) sulfide}}{FeS}$$

$$4\,Li + O_2 \longrightarrow \underset{\text{lithium oxide}}{2\,Li_2O}$$

2. The combination of two compounds to form a third:

$$CaO + CO_2 \longrightarrow \underset{\text{calcium carbonate}}{CaCO_3}$$

$$SO_2 + H_2O \longrightarrow \underset{\text{sulfurous acid}}{H_2SO_3}$$

Decomposition Reactions

Decomposition is the reverse of combination. In this case, a compound is decomposed into two or more substances, which can be either elements or other compounds.

$$AB \longrightarrow A + B$$

For example, many substances will decompose under the heat of the Bunsen burner to evolve oxygen.

$$2\,HgO \longrightarrow 2\,Hg + O_2$$

$$2\,KNO_3 \longrightarrow 2\,KNO_2 + O_2$$

Single Displacement (Replacement) Reactions

Single displacement involves the reaction of an element with a compound such that the element replaces one of the elements in the compound (setting it free) while combining with the other element.

$$A + BC \longrightarrow AC + B$$

Examples include the reaction of an active metal with a compound of another metal that is less active, or with an acid displacing hydrogen.

$$Ba + ZnSO_4 \longrightarrow BaSO_4 + Zn$$

$$Zn + HgCl_2 \longrightarrow ZnCl_2 + Hg$$

$$2\ Al + 6\ HCl \longrightarrow 2\ AlCl_3 + 3\ H_2$$

However, note that Zn will not react with $BaSO_4$ since Zn is a less active metal than Ba. Reactions such as this provide the experimental evidence for the activity or electromotive force series of the metals (see your textbook).

$$Zn + BaSO_4 \longrightarrow \text{no reaction}$$

Double Displacement (Metathesis) Reactions

The general form of the double displacement reaction is

$$AB + CD \longrightarrow AD + BC$$

Thus, in this reaction type, the positive and negative ions simply exchange partners. Such reactions occur in solution if one of three criteria is satisfied: (1) formation of a precipitate, (2) formation of a gas, or (3) formation of a weak electrolyte. If none of the possible products (AD or BC) is insoluble (precipitates), evolves as a gas, or is a weak electrolyte (such as water), then no reaction occurs.

Precipitation Reactions (Aqueous Solution)

Two compounds, which are water soluble, react to form new compounds, one of which is insoluble in water.

$$AgNO_3 + NaCl \longrightarrow AgCl(s) + NaNO_3$$

silver nitrate   sodium chloride   silver chloride   sodium nitrate

$$BaCl_2 + Na_2SO_4 \longrightarrow BaSO_4(s) + 2\ NaCl$$

barium chloride   sodium sulfate   barium sulfate   sodium chloride

In the above equations, all compounds are quite water soluble except AgCl and $BaSO_4$. A precipitate is indicated by the (s) for solid after the formula of the compound. In order to determine whether a precipitate forms, the following solubility rules can be used.

General Rules for Solubility of Ionic Compounds in Water

1. Nitrates, chlorates, and acetates are soluble.
2. Chlorides, iodides, and bromides are soluble, except those of Ag^+, Pb^{2+}, and Hg_2^{2+}.
3. Sulfates are soluble, except those of Ba^{2+}, Sr^{2+}, and Pb^{2+}. $CaSO_4$ and Ag_2SO_4 are slightly soluble.
4. Hydroxides are insoluble, except those of the alkali metals and NH_4^+. The hydroxides of alkaline earths are sparingly soluble.
5. Carbonates, phosphates, and silicates are insoluble, except those of the alkali metals and NH_4^+.
6. Sulfides are insoluble, except those of the alkali metals and NH_4^+. The sulfides of Mg^{2+}, Al^{3+}, Cr^{3+}, and the alkaline earth metals cannot be precipitated because they decompose.

Gas Formation

If a gas is evolved, a reaction will occur, as in the following examples:

$$H_2SO_4 + Na_2CO_3 \longrightarrow Na_2SO_4 + H_2O + CO_2(g)$$

$$K_2S + 2\ HCl \longrightarrow H_2S(g) + 2\ KCl$$

$$NH_4Cl + NaOH \longrightarrow NaCl + NH_3(g) + H_2O$$

Gas evolution—(g) for gas—often is a result of the reaction of an acid or base with a salt, and the gas will ordinarily be comprised of two nonmetals.

Weak Electrolyte Formation

In all acid–base neutralization reactions, the formation of water is either the driving force for the reaction or a major contributing factor. Water, a weak electrolyte, forms from H^+ and OH^-, removing these ions from solution and simultaneously giving off energy–and thereby becoming the driving force for the reaction. In this way, weak electrolyte formation is like precipitate and gas formation in removing ions from solution. Since nonmetal oxides are acids and metal oxides are bases, such substances may participate in acid–base neutralization. Some examples of hydroxides and oxides reacting follow.

1. Acid–base neutralization.

HCl	+	KOH	$\longrightarrow$	KCl	+	H_2O
hydrochloric acid (an acid)		potassium hydroxide (a base)		potassium chloride (a salt)		water
H_2SO_4	+	$Ca(OH)_2$	$\longrightarrow$	$CaSO_4$	+	$2\ H_2O$
sulfuric acid (an acid)		calcium hydroxide (a base)		calcium sulfate (a salt)		water

2. Metal oxide + acid $\longrightarrow$ salt + water. When oxides of many metals are added to water, bases are formed. Many metallic oxides are basic anhydrides and act as bases toward acids.

Na_2O	+	H_2O	$\longrightarrow$	$2\ NaOH$
sodium oxide (a basic oxide)				sodium hydroxide

The reaction between a metallic oxide and an acid may be considered to be a neutralization with the formation of a salt and water.

CaO	+	$2\ HCl$	$\longrightarrow$	$CaCl_2$	+	H_2O
calcium oxide (a basic oxide)		hydrochloric acid (an acid)		calcium chloride (a salt)		

3. Nonmetal oxide + base ⟶ salt + water. Oxides of many nonmetals are acid anhydrides. When many of the oxides of nonmetals are added to water, acids are formed. Therefore, nonmetal oxides act as acids toward a base.

$$SO_2 \quad + \quad H_2O \quad \longrightarrow \quad H_2SO_3$$

   sulfur dioxide   sulfurous acid

   (an acidic oxide)

The reaction between an oxide of a nonmetal and a base may thus be considered to be a neutralization with the formation of a salt and water.

$$SO_3 \quad + \quad 2\ KOH \quad \longrightarrow \quad K_2SO_4 \quad + \quad H_2O$$

   sulfur trioxide   potassium hydroxide   potassium sulfate

   (an acidic oxide)   (a base)   (a salt)

$$CO_2 \quad + \quad 2\ Ca(OH)_2 \quad \longrightarrow \quad CaCO_3 \quad + \quad H_2O$$

   carbon dioxide   calcium hydroxide   calcium carbonate

   (an acidic oxide)   (a base)   (a salt)

SAFETY PRECAUTIONS

Review the safety rules on pages 1 and 2.

Eye protection is always necessary, but it is especially important in this experiment.

Be particularly careful when heating and mixing chemicals. Hold test tubes with a test tube holder.

Use the hood to exhaust any fumes that might be released when carrying out the tests that follow.
Make sure that you use only small quantities of chemicals.
Reactions are particularly dangerous when you don't know what to expect—therefore, the more prepared you are to do the tests in this experiment, and the more careful you are, the less likely you are to have an accident.
Use particular care when using *30% H_2O_2*. If any gets on your skin, wash immediately with water for at least 5 minutes.

PROCEDURE

On the report, give your observation as to whether a reaction actually occurs along with your experimental evidence (evolution of heat, evolution of gas, splint test, litmus test, color change, and so on). Also, write the equation for any reaction that you predict will occur. See the completed examples below.

A. Combination Reactions

1. Test the following substances for reaction with oxygen by taking a small quantity (about the size of a pea) and holding it in the flame of the burner on a deflagrating spoon or spatula: Mg, Ca, Cu, S, P. (**Caution:** Use the hood for S and P.)

2. Mix approximately 0.1 g of the following substances with 10 drops of water in a test tube: BaO, Al_2O_3, CO_2 (dry ice). Test the resulting solution with litmus paper by dipping a clean stirring rod into the solution and then touching the stirring rod to a piece of red litmus paper. Repeat this using blue litmus paper. A base will turn red litmus paper blue. An acid will turn blue litmus paper red.

   DISPOSAL
   Copper: Rinse and put into a separate bottle for copper metal.
   S, P: Put into a waste bottle labeled *reducing agents—solids*.
   BaO, Al_2O_3: Filter solutions into waste bottle labeled *heavy metals—solutions*. Filtered solids will be put with other heavy metal waste solids.

B. Decomposition Reactions

Test the following substances for decomposition by heating approximately 0.1 g of each substance in separate test tubes: CuO; $CaCO_3$; Na_2SO_4; $KClO_3$; $KClO_3$ mixed with MnO_2; and 30% H_2O_2. Test each solid before and after heating with moist litmus paper. Perform a splint test on any gases that are produced. A splint test is used to identify gases produced in a chemical reaction. A wooden splint is ignited and then extinguished. While the splint is still glowing, it is inserted into the test tube where the reaction has occurred. If the gas produced by the reaction is oxygen the splint will momentarily glow more brightly and may even ignite. If hydrogen gas is present, there will be an audible pop as the glowing splint is inserted into the test tube. If the splint is completely extinguished, then the substance must be a nonflammable gas such as CO_2, N_2, or Cl_2.

DISPOSAL Na_2SO_4, $CaCO_3$: *Dispose* in the classroom trash container.

$\quad$ **CuO:** *Dispose* in a waste bottle labeled *heavy metals—solids*.

$\quad$ **$KClO_3$, MnO_2:** *Dispose* in a waste bottle labeled *oxidizing agents—solids*.

Single Displacement Reactions

1. Test the following substances for displacement of hydrogen in water by adding approximately 0.1 g to 10 drops of water in a test tube: Li, Ca, Fe, Zn, Al, Pb. If no reaction occurs at room temperature, heat the mixture. (**Caution:** Be particularly careful with Li.)

$\quad$ **DISPOSAL** $\quad$ **Unreacted metals:** Rinse and *recycle* into the original container, or *dispose* in a waste bottle labeled *metals*.

$\quad$ **Solutions:** Add phenolphthalein and neutralize with 1 M HCl (first loss of color). Flush down the sink, or dispose in a labeled waste bottle.

2. Test the same substances (as in Step 1) in 10 drops of dilute HCl solution.

$\quad$ **DISPOSAL of Acid solutions** $\quad$ Add phenolphthalein and neutralize with 1 M NaOH (first appearance of permanent pink color). Filter any solid into a large beaker and flush the resulting filtrate solution down the sink with running water. Put solid in a waste bottle labeled *heavy metals—solids*.

3. Mix the following materials in a test tube and observe for evidence of a reaction. Use about 0.1 g of metal and 10 drops of solution.

$\quad$ Zn + 0.1 M $CuSO_4$ $\qquad$ Al + 0.1 M $ZnCl_2$ $\qquad$ Cu + 0.1 M $FeSO_4$ $\qquad$ Ni + 0.1 M $Ca(NO_3)_2$

4. For those combinations in step 3 that did not react, mix the opposite combination. For example, if Zn does not react with 0.1 M $CuSO_4$, mix Cu with 0.1 M $ZnSO_4$.

$\quad$ **DISPOSAL** $\quad$ **SOLUTIONS AND RESIDUES:** *Filter* into a waste bottle labeled *heavy metals—solutions*. Solid wastes will be put into a waste bottle labeled *metals*.

Double Displacement (Metathesis) Reactions

Mix solutions of the following pairs of substances and observe for evidence of reaction (all solutions are 0.1 M). Use 2 drops of each solution on a spot plate.

$AgNO_3$ + NaCl $\qquad$ NaOH + CO_2 $\qquad$ CaO + H_2SO_4 $\qquad$ $Pb(NO_3)_2$ + KI

NH_4Cl + NaOH $\qquad$ $CuSO_4$ + $Ni(NO_3)_2$ $\qquad$ $BaCl_2$ + H_2SO_4 $\qquad$ Na_2CO_3 + H_2SO_4

DISPOSAL $\quad$ Treat the same as in Part C, step 4, except for residues from **NH_4Cl + NaOH, NaOH + CO_2, and Na_2CO_3 + H_2SO_4:** Add phenolphthalein and *neutralize* with 1 M HCl (the first disappearance of pink color) or 1 M NaOH (first appearance of permanent pink color). Flush the resulting salt solution down the sink with running water.

Prelab   Name _____ Section _____

Types of Chemical Reactions

1. What is the safety precaution associated with any gaseous fumes produced by a reaction in this experiment?

2. In each of the parts of the procedure you are combining chemicals and observing whether a chemical reaction occurs. What are some observations that will indicate that a reaction has occurred?

3. What test is used to determine the identity of gases produced during a reaction.

4. Classify each of the following reactions into one of the four general categories: combination, decomposition, single displacement, or double displacement.

   a. $H_2SO_4(aq) + 2\ KOH(aq) \rightarrow K_2SO_4(aq) + 2\ H_2O(l)$

   b. $2\ S(s) + 3\ O_2(g) \rightarrow 2\ SO_3(g)$

   c. $NH_4NO_2(s) \rightarrow N_2(g) + 2\ H_2O(g)$

   d. $2\ Al(s) + 6\ HCl(aq) \rightarrow 2\ AlCl_3(aq) + 3\ H_2(g)$

Report
Types of Chemical Reactions

Name _____ Section _____

A. Combination Reactions

Reaction	Prediction	Observation	Equation
1. $Li + O_2 \rightarrow$?	Yes, Li reacts with O_2	A white coating forms on the Li	$4\,Li + O_2 \rightarrow 2\,Li_2O$
$Mg + O_2 \rightarrow$?			
$Ca + O_2 \rightarrow$?			
$Cu + O_2 \rightarrow$?			
$S + O_2 \rightarrow$?			
$P + O_2 \rightarrow$?			
2. $CaO + H_2O \rightarrow$?	Yes, CaO reacts with H_2O	Solution gets warm; turns litmus paper blue	$CaO + H_2O \rightarrow Ca(OH)_2$
$BaO + H_2O \rightarrow$?			
$Al_2O_3 + H_2O \rightarrow$?			
$CO_2 + H_2O \rightarrow$?			

B. Decomposition Reactions

Reaction	Prediction	Observation	Equation
$KNO_3 \rightarrow$?	Yes, KNO_3 decomposes on heating	A gas is evolved; a glowing splint ignites	$2\,KNO_3 \rightarrow 2\,KNO_2 + O_2$
$CuO \rightarrow$?			
$CaCO_3 \rightarrow$?			
$Na_2SO_4 \rightarrow$?			
$KClO_3 \rightarrow$?			
$KClO_3$ and $MnO_2 \rightarrow$?			
$H_2O_2 \rightarrow$?			

Report   Name _____ Section _____

C. Displacement Reactions

	Reaction	Prediction	Observation	Net Ionic Equation
1.	$Na + H_2O \rightarrow$?	Na reacts with H_2O	Rapid & violent reaction occurs; gas evolves; a glowing splint causes a pop	$2\,Na + 2\,H_2O \rightarrow 2\,Na^+OH^- + H_2$
	$Li + H_2O \rightarrow$?			
	$Ca + H_2O \rightarrow$?			
	$Fe + H_2O \rightarrow$?			
	$Zn + H_2O \rightarrow$?			
	$Al + H_2O \rightarrow$?			
	$Pb + H_2O \rightarrow$?			
2.	$Mg + HCl \rightarrow$?	Mg reacts with HCl	Vigorous reaction occurs; rapid evolution of gas; a glowing splint causes a pop	$Mg + 2\,H^+ \rightarrow Mg^{2+} + H_2$
	$Li + HCl \rightarrow$?			
	$Ca + HCl \rightarrow$?			
	$Fe + HCl \rightarrow$?			
	$Zn + HCl \rightarrow$?			
	$Al + HCl \rightarrow$?			
	$Pb + HCl \rightarrow$?			

Report   Name _____ Section _____

Reaction	Prediction	Observation	Net Ionic Equation
3. $Ca + CuSO_4 \rightarrow$?	Ca reacts with $CuSO_4$	Solution decolorizes; copper precipitates	$Ca + Cu^{2+} \rightarrow Ca^{2+} + Cu^0$
$Cu + CaSO_4 \rightarrow$?	No reaction	No result	No reaction
$Zn + CuSO_4 \rightarrow$?			
$Al + ZnCl_2 \rightarrow$?			
$Cu + FeSO_4 \rightarrow$?			
$Ni + Ca(NO_3)_2 \rightarrow$?			
4. $Cu + ZnSO_4 \rightarrow$?			
$Zn + AlCl_3 \rightarrow$?			
$Fe + CuSO_4 \rightarrow$?			
$Ca + Ni(NO_3)_2 \rightarrow$?			

D. Double Displacement Reactions

Reaction	Prediction	Observation	Net Ionic Equation
$Hg_2(NO_3)_2 + H_2SO_4 \rightarrow$?	Reaction Occurs	A white precipitate forms	$Hg_2^{2+} + SO_4^{2-} \rightarrow Hg_2SO_4$
$AgNO_3 + NaCl \rightarrow$?			
$NaOH + CO_2 \rightarrow$?			
$CaO + H_2SO_4 \rightarrow$?			
$Pb(NO_3)_2 + KI \rightarrow$?			
$NH_4Cl + NaOH \rightarrow$?			
$CuSO_4 + Ni(NO_3)_2 \rightarrow$?			
$BaCl_2 + H_2SO_4 \rightarrow$?			
$Na_2CO_3 + H_2SO_4 \rightarrow$?			

Report   Name _____ Section _____

QUESTIONS

1. Give experimental evidence to support the identity of each of the gaseous products in Part B of the experiment.

2. From your observations, what is the order of activity of the metals in Part C?

3. Identify which of the three driving forces (formation of a precipitate, a gas, or a weak electrolyte) was associated with each of the reactions in Part D.

4. From the information available in the introduction to this experiment or from your textbook or other sources, predict the products of each of the following reactions. Assume that a reaction occurs in each case. Write a balanced equation for each one.

    a. $Mg(s) + Zn(NO_3)_2(aq) \rightarrow$

    b. $CuBr_2(aq) + Na_2S(aq) \rightarrow$

    c. $Sr(s) + O_2(g) \rightarrow$

    d. $Al_2O_3(s) \rightarrow$

    e. $Na(s) + CuCl_2(aq) \rightarrow$

5 The Stoichiometry of a Reaction

INTRODUCTION

The object of this experiment is to determine the mole ratio with which two substances combine chemically. Such information is helpful in determining the stoichiometry of the reaction – that is, the balanced chemical equation and the information that can be derived from it.

The principle underlying this experiment is based on maintaining a constant quantity of reactant A while varying the quantity of reactant B in a series of experiments. The mass of an insoluble product C is then used as an indicator of the effect of varying the quantity of B. When sufficient B has been added to react completely with the constant quantity of A there will be no further increase in C and the mole ratio of A to B can be deduced.

For example, the stoichiometry of the cadmium nitrate–sodium sulfide system was investigated using this technique. With a constant volume (6.00 mL) of 1.00 M $Cd(NO_3)_2$ solution [6.00 mmol $Cd(NO_3)_2$], the quantity of insoluble product was plotted against the number of millimoles of Na_2S (Figure 5-1). The mass of precipitate formed for each volume of the Na_2S solution is given in Table 5-1. The two straight lines in Figure 5-1 intersect at 0.85 g of precipitate, which corresponds to 6.00 mmol Na_2S. Since 6.00 mmol of $Cd(NO_3)_2$ was used, the mole ratio of Na_2S to $Cd(NO_3)_2$ is 6.00 to 6.00 or 1:1. This suggests that the precipitate could be CdS or $NaNO_3$. Since nitrates are generally soluble and sulfides are generally insoluble, CdS is indicated as the precipitate. The equation for the reaction can then be written.

$$Na_2S(aq) + Cd(NO_3)_2(aq) \longrightarrow CdS(s) + 2\ NaNO_3(aq)$$

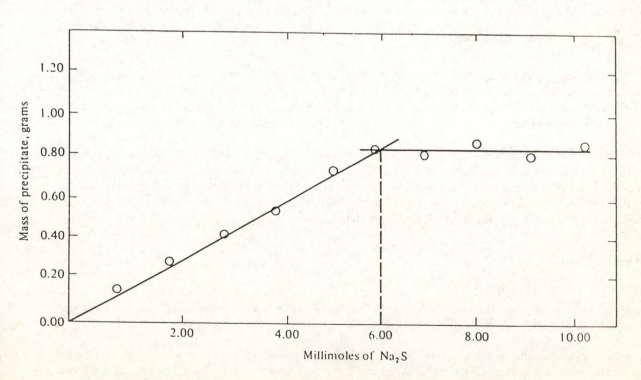

Figure 5-1. Plot of Mass of Precipitate versus Millimoles of Na₂S

51

Table 5-1. Data for Reaction of 1.00 M Na_2S with 1.00 M $Cd(NO_3)_2$

Assignment Number	1.000 M Na_2S		1.00 M $Cd(NO_3)_2$		Precipitate, g
	Volume, mL	Number of mmol	Volume, mL	Number of mmol	
1	1.00	1.00	6.00	6.00	0.145
2	2.00	2.00	6.00	6.00	0.285
3	3.00	3.00	6.00	6.00	0.438
4	4.00	4.00	6.00	6.00	0.590
5	5.00	5.00	6.00	6.00	0.740
6	6.00	6.00	6.00	6.00	0.860
7	7.00	7.00	6.00	6.00	0.865
8	8.00	8.00	6.00	6.00	0.879
9	9.00	9.00	6.00	6.00	0.869
10	10.00	10.00	6.00	6.00	0.867

To confirm the identity of the precipitate, one computes the masses of $NaNO_3$ and of CdS that would be formed from 6.00 mmol Na_2S.

$$6.00 \text{ mmol } Na_2S \times \frac{2 \text{ mmol } NaNO_3}{1 \text{ mmol } Na_2S} \times \frac{0.085g \text{ } NaNO_3}{1 \text{ mmol } NaNO_3} = 1.02 \text{ g } NaNO_3$$

$$6.00 \text{ mmol } Na_2S \times \frac{1 \text{ mmol } CdS}{1 \text{ mmol } Na_2S} \times \frac{0.145g \text{ } CdS}{1 \text{ mmol } CdS} = 0.870 \text{ g } CdS$$

Since the constant mass of precipitate averages 0.868 g, which is much closer to the 0.870 mass value of CdS than to the 1.02 mass value of $NaNO_3$, the identity of the precipitate as CdS is confirmed.

FORMING PRECIPITATES

When a solid precipitate forms from a supersaturated solution, the particle size may be so small that some of the particles pass through the filter paper. This is the case with the bright yellow precipitate formed in this experiment. To create larger particles of solid, nitric acid is added to the reaction mixture and the mixture is gently heated.

Nitric acid neutralizes the excess charge on the surface of the precipitate particles. This excess charge consists of extra anions or cations on the surface of each particle. For example, when CdS(s) forms from a solution with excess Cd^{2+}(aq) ion, the particles of solid are positively charged. Since like charges repel, the solid particles cannot coagulate into larger particles until the excess positive charge is neutralized.

Heating the reaction mixture dissolves the smaller particles of precipitate. This works because most solids are more soluble at high temperatures and because dissolving occurs at the surface of a particle. The smaller particles have a relatively larger surface area than the larger ones. When the heated solution is allowed to cool, the dissolved solid reprecipitates on the larger particles that remain. Thus the net result of this technique, known as digestion, is to replace the small particles with large ones.

SAFETY PRECAUTIONS

Review the safety rules on pages 1 and 2.

Be very careful in handling chemicals. Wipe up all spills promptly with a damp sponge or towel. If any chemicals get on your hands or clothing, wash immediately with soap and running water.

Heat the reaction mixture on a hot plate in the hood with the glass door closed as much as possible. Watch the reaction mixture constantly while it is being heated and stir it continually to minimize spattering. Do not look down into the beaker while it is being heated. Be very careful when handling hot beakers; use either beaker tongs or a carefully folded towel.

6 Identification of Common Chemicals

INTRODUCTION

The main purpose of this experiment is to familiarize yourself with some of the properties of common chemicals. In addition, you will gain experience with several physical and chemical methods that are commonly used in chemical analysis. Finally, by developing your own scheme for identifying the various chemicals, you will begin to learn how to plan an analysis logically and will gain some appreciation for the method of scientific investigation.

Your objective for this experiment is to develop a series of tests that will allow you to identify the chemicals listed in Table 6-1. You must find properties for each chemical that allow you to distinguish it from all of the others in that table. This experiment is divided into two parts. In Part A, an assortment of chemicals and methods of identification are examined. Part B is a timed test to be taken by all students in each laboratory section simultaneously. Part A may begin very early in the course, perhaps at the end of another experiment with an introduction by your instructor to some of the tests that can be used in identification. Part B will then be scheduled toward the end of the course. If this is the case, you can use extra time at the end of other experiments to test some of the chemicals. If you keep careful records of your results in your laboratory notebook, you will be able to devise a complete scheme for the identification of any of the chemicals as soon as you have tested the last of them. Then you only need to try out your scheme to see that it works well.

SAFETY PRECAUTIONS

Review the safety rules on pages 1 and 2.

Wear goggles at all times.

Never taste any chemical.

Smell chemicals by fanning some of the vapors toward your nose and then only when specifically directed to do so by your instructor. **Never** smell a reaction mixture when it is being heated or when it is reacting.

If you must touch chemicals or if you inadvertently get chemicals on you, wash immediately with soap and large volumes of running water. Report any contact with chemicals to your instructor.

Use small quantities of chemicals as directed. Compare the amounts you take with those in labeled test tubes prepared by your instructor or weigh the chemicals.

Stay at a distance from a reaction mixture while it is being heated.

Never use anything but a glass stirring rod to mix chemicals. If you can, mix chemicals by gently tapping the side of a test tube with your finger. (Have your instructor demonstrate this technique.)

Never bring large quantities of liquid near an open flame.

Dispense reagents either with the dropper or spatula supplied with each bottle or by pouring out a small quantity of liquid or solid reagent. No more than 0.5 mL (10 drops) of liquid or 0.3 grams (size of 3 grains of rice) of solid should be taken back to your bench for testing. Chemicals should be dispensed into your smallest sized test tubes, not into flasks or beakers.

Do not return excess reagent to a bottle; it may contaminate the rest of the contents. Instead, throw it away in the manner and place designated by your instructor, normally in the appropriate waste container.

PROCEDURE

Part A

1. On the reagent shelf you will find labeled bottles of the different chemicals listed in Table 6-1. Your task is to devise a series of tests that will enable you to distinguish between the various chemicals.
2. Some tests will be quite simple, such as looking at a substance and observing its color and texture. You may smell the chemicals if you have your instructor's explicit permission to do so. But, do so cautiously by opening the bottle and fanning some of the vapors toward your nose. Do not taste the chemicals! Since many chemicals are similar in appearance, you would do well to devise a confirming test.
3. All the chemical tests should be performed with a drop of liquid or a piece of solid no larger than a match head. Larger quantities will waste chemicals, yield ambiguous results, and may produce dangerous reactions.
4. You should begin by looking up the physical and chemical properties of the substances; use your textbook or the library. Two good reference books are the *Handbook of Chemistry and Physics* and *The Merck Index.* You may also use the results of your laboratory work throughout the semester.
5. Some of the tests that you might wish to use follow. Try these tests in the order listed until you find a test that gives definite results.
    a. Rate of evaporation.
    b. Viscosity. How readily does a liquid flow?
    c. Solubility. Make sure that the solid is finely divided and try to dissolve an amount of solid the size of a match head in 0.5 mL of water.
    d. Color of a dilute aqueous solution.
    e. Effect of aqueous solutions or pure liquids on litmus paper. Acidic solutions turn blue litmus red. Basic solutions turn red litmus blue.
    f. Reaction with precipitating agents, such as $BaCl_2$(aq) and $AgNO_3$(aq). Test 1 drop of your aqueous solution on a spot plate with 1 drop of precipitating agent.
    g. Solubility in solvents other than water, such as ethanol.
    h. Ability to decolorize a dilute solution of $KMnO_4$ (reducing power).
    i. Reactions with a strong (NaOH) and a weak [NH_3(aq)] base.
    j. Reactions with a strong (HCl) and a weak ($HC_2H_3O_2$) acid.
    k. Flammability and/or color of the flame. Test one drop of the liquid or the aqueous solution held on a loop of nichrome wire.
6. *Precautions*
    a. Some of the chemicals are highly poisonous. **Do not taste any of the chemicals.**
    b. Many of the liquids are highly flammable. **Do not bring large quantities of any liquid near an open flame.**
7. The most convenient time for you to work will be at the end of a laboratory period after you have completed another experiment but have some time left. Consult your instructor for other times when you may be allowed to work in the laboratory.
8. Now, after you have made your observations, carefully plan your tests. Record all information in your notebook. You will only be allowed the use of your notebook during the testing period.
9. Try out your tests by having another student select several chemicals for you to identify. Modify your tests if necessary.

Some Suggestions for Tests

Your investigation of the common chemicals is best done in a systematic fashion. Indeed, part of the purpose of this laboratory is to develop your skill in designing and carrying out an experiment. You should keep in mind that in Part B you will have only a limited amount of time to test each chemical (4 min) and therefore, your procedures should be as simple as possible. There are two consequences of this: First, you do not want to perform more tests than are absolutely necessary. Thus, you should ask yourself the question: Does this last test uniquely identify this chemical? Discard any tests that don't contribute more knowledge about the chemical. Second, since physical tests (color, liquid or solid, viscosity, etc.) are generally easier and faster than chemical tests, it is to your advantage to devise as many physical tests as you can.

Specific Suggestions

1. Start simply. Note everything that you can about a substance before you open the bottle, including answers to the following questions.
   a. Is it a solid or a liquid?
   b. What is its color?
   c. If it is a liquid, does it flow easily? That is, is it viscous or not?
   d. If it is a solid, is it shiny or dull? How large are the particles?
2. Open the bottle and take out a match head sized piece of solid or 5 drops of liquid. Here again you can make a number of observations.
   a. What does it smell like? Do this with caution by waving some of the vapors toward your nose, not by snorting the vapors directly over the bottle. Some of the smells (ammonia and acetic acid, for example) are so characteristic that you can stop here. Do not forget to smell the solids also. But do not smell anything unless specifically directed to do so by your instructor.
   b. If it is a liquid, does it have a high (a drop beads up on a glass plate) or a low (a drop spreads out) surface tension ?
   c. If it is a liquid, is it volatile (the drop evaporates rapidly) or nonvolatile? High surface tension and low volatility usually go together.
   d. If it is a solid, does it crush easily? (What is its tensile strength?) Try to crush it with the end of a glass rod. Does it look different when it is crushed?
3. Try dissolving about 0.1 g of the crushed solid in 0.5 mL of distilled water in a small test tube using a stirring rod to agitate the mixture.
   a. Does it dissolve easily?
   b. What is the color of the solution?
   c. Did any gas evolve, did you hear anything, or was any heat produced when you dissolved it in water? In other words, did a chemical reaction occur with water?
   d. Does the freshly prepared solution have a smell? Get your instructor's permission before you smell a solution.
4. Test the solution or the liquid sample chemically. Chemical reactions are best done using one drop of reagent on a spot plate.
   a. Is it acidic, basic, or neutral? Dip the stirring rod into the solution and touch it to a piece of litmus paper. Using this method, one piece of litmus paper can be used to test several solutions. If it turns blue litmus red, the solution is acidic; if it turns red litmus blue, the solution is basic. If no change occurs in either case, the solution is neutral.
   b. Add a drop of possible precipitating agent to one drop of liquid sample on a glass plate and see if a precipitate appears. Possible agents are dilute H_2SO_4, dilute NaOH, dilute HCl, dilute $AgNO_3$, and dilute $BaCl_2$.
      (1) Did a precipitate appear? What color is it?
      (2) Did a reaction occur and perhaps change the color of the solution (what is the new color?) or produce a gas (what does the gas smell like?)? (Before you smell the gas, get your instructor's permission.)
   c. Add a drop of dilute $KMnO_4$ solution (light purple) to 3 drops of the solution. Does the $KMnO_4$ turn colorless? Is any gas evolved? What is the new color of the solution?

d.   Try a flame test. Put a drop of solution on a loop of wire and carefully evaporate the liquid in the coolest part of the Bunsen flame. Repeat this twice. You now have a concentrated solution or some solid on the wire loop. Place this in the hottest part of the Bunsen flame and observe the brief flash of color. This is called a flame test. It is difficult to perform correctly without practice. (Many people incorrectly substitute a chunk of solid for the drops of solution. This shortcut often gives a misleading color to the flame and also tends to ruin the Bunsen burner when the melted solid drops down inside.)

5.   Only if the solid does not dissolve in water, try to dissolve it in another solvent. First try dilute acids (HCl, HNO₃, H₂SO₄) and bases (NH₃, NaOH) and then concentrated acids and bases if necessary. All the tests that you can perform on the aqueous solution (see 3 and 4 above) are valid on the solution produced at this point.

6.   In summary, if you are careful you can avoid several errors. Always work with small quantities of chemicals. Too much material is not only wasteful but also produces poor results. Practice the tests a couple of times and make sure you get the same results each time. Make sure that you have at least two positive tests for each identification. Don't distinguish between two substances on the basis of only one test, especially if the results are difficult to interpret.

Part B

A selection of 16 chemicals from those studied will be placed in numbered bottles. The timed test will be given during the last laboratory period for each section. You will be permitted the use of only your laboratory notebook during the test.

The test will be given during the first part of the laboratory period. The first 32 min will consist of eight 4-min test periods. There will then be a 5-min break. This will be followed by a second set of eight 4-min test periods and a 5-min break. At the end of each test period, a buzzer will sound and you must pass your bottle along in the indicated direction. The 5-min breaks may be used for cleaning glassware and making final tests on samples of compounds that you have saved. At the end of the entire test, there will be another 5-min break during which you are expected to complete your work and hand in your answer sheet. All papers must be submitted within 10 min of the last test period (the two five-minute breaks).

If one sample is quite easy to identify, you may not go on to the next sample, but you may go back and work on other samples that you have saved.

You are encouraged to assemble the necessary testing materials at your desk before the beginning of the timed test.

Disposal

Unreacted metals: Rinse and *recycle* in a bottle labeled for that metal.
Solids: (1) NaCl, NH₄Cl, CaCl₂, CaO, SiO₂, NaHCO₃, CaCO₃, K₂CO₃, CaSO₄, borax, and sucrose may be *disposed* directly into the room trash container. (2) All other solids should be disposed in a waste bottle labeled *common chemicals—solid wastes.*
Solid/solution mixtures: Filter into a waste bottle labeled *common chemicals—aqueous wastes.* The solids will be put into the solid waste bottle.
Aqueous solutions: Solutions of the salts listed in solids (1) above, acetic acid, acetone, alcohols, and ethylene glycol may be rinsed down the sink with running water. Other aqueous solutions must be put in a waste bottle labeled *common chemicals—aqueous wastes.*

Example: A Scheme to Identify Five White Solids

A student is to find tests that distinguish among five white solids: $Ba(NO_3)_2$, $BaCl_2$, KNO_3, NH_4NO_3, $C_7H_6O_2$ (benzoic acid). Any useful test must allow the substances being tested to be divided into at least two groups based on the results of the test. Therefore, color and state of matter are not useful properties since all five of the compounds are white solids.

Proposed Procedure:
1. Test the solubility in water. <u>Result</u>: All are quite soluble, except benzoic acid; therefore, benzoic acid is distinguished from the others.
2. Of the remaining four solids, three are nitrate salts and one is a chloride salt. Therefore, a test to distinguish chloride from nitrate is appropriate. Recall from the solubility rules that Ag^+ ions form a precipitate when mixed with Cl^- ions. A small amount of each solid is dissolved in 5 drops of water. One drop of $AgNO_3$ is added to one drop of each of the four solutions. <u>Result</u>: A white precipitate of AgCl is observed in the solution of $BaCl_2$.
3. The three remaining solids all have nitrate as the anion. Therefore, they must be distinguished based on the properties of the cation. Recall from the solubility rules that $BaSO_4$ is insoluble. One drop of H_2SO_4 is added to one drop of each of the three solutions. <u>Result</u>: A white precipitate of $BaSO_4$ is observed in the $Ba(NO_3)_2$ solution.
4. Solutions of the two remaining solids can be tested for acidity using litmus paper. <u>Result</u>: NH_4NO_3 is acidic and KNO_3 is neutral.
5. The flow chart in Figure 6-1 summarizes the procedure used to identify the five solids. It can serve as a quick reference during a test.

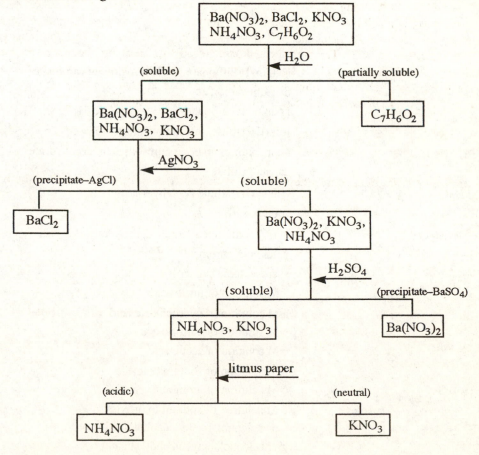

Figure 6-1.   Flow Chart of Proposed Procedure

Table 6-1. Study Set of Common Chemicals

Mg	magnesium metal
Al	aluminum metal
Zn	zinc metal
Sn	tin metal
Fe	iron metal
C	carbon, solid
S	sulfur, solid
Fe_2O_3	iron(III) oxide, ferric oxide
MnO_2	manganese(IV) oxide, manganese dioxide
$KMnO_4$	potassium permanganate
$CuBr_2$	copper(II) bromide, cupric bromide
NaCl	sodium chloride
NH_4Cl	ammonium chloride
$CaCl_2$	calcium chloride
$SrCl_2$	strontium chloride
$Cu(NO_3)_2 \cdot 3H_2O$	copper(II) nitrate trihydrate, cupric nitrate trihydrate
$Ni(NO_3)_2 \cdot 6H_2O$	nickel(II) nitrate hexahydrate
ZnO	zinc oxide
CaO	calcium oxide
SiO_2	silicon dioxide (sand)
$NaHCO_3$	sodium bicarbonate, sodium hydrogen carbonate
$CaCO_3$	calcium carbonate
K_2CO_3	potassium carbonate
FeS	iron(II) sulfide, ferrous sulfide
$KAl(SO_4)_2 \cdot 12H_2O$	alum, aluminum potassium sulfate dodecahydrate
$CaSO_4$	calcium sulfate
$CuSO_4 \cdot 5H_2O$	copper(II) sulfate pentahydrate
Na_2SO_3	sodium sulfite
$Na_2B_4O_7 \cdot 10H_2O$	borax, sodium tetraborate decahydrate
K_2CrO_4	potassium chromate
$C_{12}H_{22}O_{11}$	sucrose, table sugar
HNO_3	3 M solution of nitric acid
HCl	6 M solution of hydrochloric acid
H_2SO_4	6 M solution of sulfuric acid
$HC_2H_3O_2$	6 M solution of acetic acid
NaCl	3 M solution of sodium chloride
$NH_3(aq)$	6 M solution of ammonia
NaOH	6 M solution of sodium hydroxide
CH_3OH	wood alcohol, methyl alcohol, methanol
C_2H_5OH	grain alcohol, ethyl alcohol, ethanol
$(CH_3)_2CO$	acetone
C_6H_{12}	cyclohexane
$C_2H_4(OH)_2$	ethylene glycol, 1,2-ethanediol

7 Titration of Acids and Bases

INTRODUCTION

This experiment is one that will already be familiar to many students. For several reasons, however, it might be worthwhile to repeat it. First, it gives us a chance to improve and perfect our experimental technique, especially for the degree of accuracy and precision required in this exercise. Second, it can serve to reinforce our knowledge of stoichiometry as well as to reinforce the self–confidence that comes with doing a thing well and with understanding.

Volumetric titration is the process of mixing measured volumes of reacting solutions in such a manner that you can determine when chemically equivalent amounts of reactants are present. The purpose of titration is to determine the concentration of a solution of unknown strength. The concentration of one of the solutions, expressed as molarity in this case, must be known. After titration, the molarity of the other solution can be calculated from the measured data.

The *equivalence point* of a titration is the point at which equivalent amounts of reactants have been mixed. In order to determine the equivalence point, a visual *indicator* usually is added to the solution to be titrated. Such an indicator, if properly selected, undergoes a sharp color change at (or very near) the equivalence point. The point of the titration at which the indicator changes color is known as the *end point*. If the indicator is chosen correctly, the endpoint will occur at the same time when the equivalence point has been reached.

In acid–base titrations, phenolphthalein often is used as the indicator (Figure 7-1). Phenolphthalein is colorless in acidic solution. At a pH of about 8.3, it undergoes a sharp change to a pink color as you add a base. At a pH of about 10, it is red.

Other visual indicators are available for different pH ranges. In the titration of a strong acid with a strong base, the pH of the solution increases rapidly and has a value of 7 at the equivalence point. Since equivalence points for many titrations lie in the range of pH from 7 to 9, phenolphthalein is an excellent general purpose acid–base indicator.

SAFETY PRECAUTIONS

Review the safety rules on pages 1 and 2.

Both HCl and NaOH are harmful to skin and eyes. If any gets on your skin, wash thoroughly with water.

Wear eye protection! If any gets in your eye, rinse immediately and thoroughly with water.

Never add water to acid. Always add acid to water.

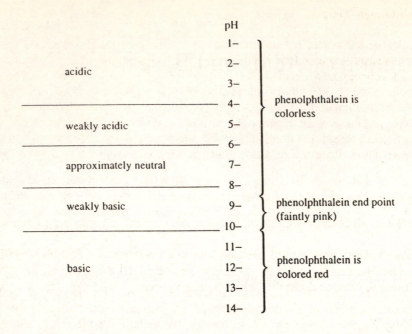

Figure 7-1. Phenolphthalein Behavior Over the pH Scale

PROCEDURE

A. Standardization of NaOH Solution

Into a clean 500 mL Florence flask place enough NaOH pellets to provide about 2 g of compound. Add approximately 250 mL of water and mix thoroughly to dissolve the NaOH. Add 150 mL more water and mix thoroughly. Add water to the bottom of the neck of the flask and mix thoroughly. This should yield a homogeneous solution of approximately 0.1 M NaOH. The concentration of your NaOH solution is now known only approximately, and it will have to be standardized in order to know it to 4 significant figures.

You will be supplied with a clean buret and a clean pipet. Rinse your buret twice with about 5 mL of your 0.1 M NaOH solution. Next, fill the buret to near the 0.00 mL mark with this solution. Make sure the tip of the buret is filled and contains no air. The level of the liquid in the buret is read at the bottom of the meniscus.

Rinse the pipet twice with about 5 mL of the standard 0.1000 M HCl solution. This solution will be available in the laboratory. Transfer 25.00 mL of the standard 0.1000 M HCl solution into a clean Erlenmeyer flask. Add 2 or 3 drops of phenolphthalein indicator solution. Titrate this acid solution with your NaOH solution until a faint pink color persists for over 15 sec (See Note). Swirl the solution in the Erlenmeyer flask as you titrate and rinse the walls of the flask with distilled water from a wash bottle prior to the end point. Record the volume (to the nearest ±0.01 mL) of base used.

Repeat the titration at least twice using 25.00 mL portions of 0.1000 M HCl solution and record your data. The molarities of your base as calculated from the titration data should agree to within 1% (relative average deviation) of the average value. If this is not the case, repeat the titration with 25.00 mL of 0.1000 M HCl until two successive trials agree to within 1%.

Note: The color will fade on sitting as the carbon dioxide gas in the air dissolves and reacts with the slight excess of NaOH, causing the pH to fall below 8.3.

8 Gravimetric and Volumetric Analysis

INTRODUCTION

The techniques of classical chemical analysis can be divided into two broad categories: (1) gravimetric analysis and (2) volumetric analysis. The purpose of this experiment is to employ both methods of analysis to determine the concentration of H_2SO_4 in solutions that have been prepared by dilution from a sulfuric acid solution of definite, but unknown, concentration. From the dilutions the concentrations of the unknown H_2SO_4 solution may then be calculated.

In **gravimetric analysis,** the element to be determined is isolated in a solid compound of known identity and definite composition. From the mass of this compound, the analyst can determine the mass of the element that was present in the original sample. On the other hand, when the goal is to determine the amount of a compound present in the original sample (rather than the amount of an individual element), either one of the ions or one of the elements present in that compound is isolated in a compound to be weighed. In this experiment, the amount of sulfuric acid present is determined gravimetrically by producing and weighing $BaSO_4(s)$, by reaction with $BaCl_2(aq)$.

$$H_2SO_4\ (aq) + BaCl_2(aq) \rightarrow\ BaSO_4(s) + 2\ HCl(aq) \qquad [8.1]$$

$BaSO_4$ is suitable for this gravimetric determination because it is precipitated readily and completely from the solution being analyzed. Furthermore, the resulting precipitate is pure and easily filtered, it is a solid of known and definite composition, and its molar mass is sufficiently large that an appreciable and easily weighed mass of precipitate will be produced.

In **volumetric analysis,** the amount of the species present in the sample is determined by the technique of titration. The sample is present in a solution of known volume. To this solution of unknown concentration is added another solution, called the **titrant,** of known concentration. The titrant is added in such a way that the volume added at each point is readily determined. The solute in the titrant solution is a substance that reacts quickly and completely with the species to be determined. It is also helpful if an easily observable change occurs at the **equivalence point** of the titration, the point at which the amount of titrant solute is chemically equivalent to the amount of the species being determined. If such a change does not occur, often an **indicator** is added to the solution being titrated. As a result of this addition, a color change occurs in the reaction mixture at a point in the titration very near to (or ideally at) the equivalence point. This point of color change is known as the **end point.** In this experiment, the amount of H_2SO_4 is determined volumetrically by its titration with sodium hydroxide solution.

$$H_2SO_4\ (aq) + 2\ NaOH(aq)\ \rightarrow\ Na_2SO_4(aq) + 2\ H_2O \qquad [8.2]$$

Since no perceptible change occurs at the equivalence point of this reaction (when 2 moles of NaOH have been added for every mole of H_2SO_4 originally present), phenolphthalein is used as an indicator. Phenolphthalein is colorless in acidic solution and pink in alkaline solution.

SAFETY PRECAUTIONS

Review the safety rules on pages 1 and 2 of the laboratory manual.

The original, concentrated H_2SO_4 solution can cause chemical burns if it comes in contact with your skin. Any spills or splashes should be promptly cleaned up so that others in the laboratory are not accidentally exposed. If you come in contact with this solution, you should wash the affected area for at least five minutes with running water, inform your laboratory instructor, and seek medical attention. An accidental exposure to concentrated H_2SO_4 will manifest itself after some time as a burning sensation on your skin. If you feel this sensation, **seek immediate attention.** Delay can lead to a serious skin burn.

The concentrated acid should be added to water, rather than water to the concentrated acid. ("Acid to water is the way that you oughter." Splattering or spitting often occurs when water is added to a concentrated acid because of the substantial heat of dilution that is generated.) Of course, you should wear goggles throughout this procedure (as well as during the rest of the experiment) and avoid looking down into the reaction mixture, to avoid inadvertent splashes to your hands or face.

Barium ion is a heavy metal poison and should be handled with care. Although there is no conclusive evidence that poisoning occurs on skin contact, all spills should be cleaned up promptly in order to avoid accidental contact. Once it is converted to barium sulfate, however, barium ion poses little hazard because of the extreme insolubility of the solid. Thus, glassware that has been used to hold $BaCl_2$(aq) should be rinsed with a small volume of dilute H_2SO_4. This rinse should be disposed of in the manner described for solutions in the disposal section of Part B.

PROCEDURE

A. Dilution of Concentrated H_2SO_4

1. Obtain 25 mL of 6 M H_2SO_4 in a 100 mL or smaller graduated beaker. Label this as the **initial solution.** The beaker should be scrupulously clean to avoid contamination. After use, carefully rinse the beaker with water and discard the rinsings (see the disposal section at the end of Part A).

2. Pour approximately 50 mL of distilled water in a clean 100 mL volumetric flask. Pipet exactly 10.00 mL of 6 M H_2SO_4 into the flask. (Be sure to rinse the pipet with a small volume of the sulfuric acid solution first. If you are unsure of the proper, safe procedure for pipetting, consult your instructor before beginning.) Follow the safety precautions noted above. Add distilled water to the calibration mark of the volumetric flask. Stopper securely and invert ten times to ensure thorough mixing and uniform concentration throughout the solution. This solution will be referred to as the **stock solution.** (It has a concentration of approximately 0.6 M.) It can be transferred to a clean and dry Erlenmeyer flask and stoppered for storage (and labeled *"Stock Solution"*). (Retain this stock solution until you have completed the experiment, so that you can make up more final solution if needed.)

3. Rinse the volumetric flask with several portions of distilled water. Discard the rinse water.

4. Pipet 25.00 mL of the stock solution into the 100 mL volumetric flask. This flask must be clean but not necessarily dry; you will be adding distilled water to its contents. Add distilled water, swirling it in order to mix thoroughly after each addition, to fill the flask to the calibration mark. Stopper securely and invert ten times to ensure thorough mixing. This **final solution** has a concentration of approximately 0.15 M. It should be stored in a stoppered, clean, and dry 125 mL Erlenmeyer flask (labeled *"Final Solution"*) for further use. (*Careful! Pipetting from a volumetric flask can be hazardous.*)

5. **DISPOSAL**

    Excess acid: Add phenolphthalein and neutralize with 1 M NaOH (first appearance of permanent pink color). Flush the resulting salt solution down the sink with running water.

B. Gravimetric Determination of Sulfate Ion

1. Obtain approximately 12 mL of 0.2 M $BaCl_2$(aq) in a clean, dry, stoppered 125 mL Erlenmeyer flask.

2. Determine the mass of a piece of filter paper to the nearest mg (±0.001 g). With pencil (not ink!) write an identifying mark on the outside edge of the filter paper. Fold the filter paper in quarters, make a cone, and place it in a funnel. Moisten it with distilled water so that it adheres to the walls of the funnel.

3. Using a 10 mL graduated cylinder, transfer 5.0 mL of the 0.2 M $BaCl_2$ obtained in step 1 into a 50 mL beaker. This provides a slight excess of barium ion. Warm this solution to near boiling (about 15 minutes on a hot plate). Do not heat to dryness.

4. While the above solution is heating, pipet 5.00 mL of the **final solution** of H_2SO_4 into a 50 mL beaker. Add 5 mL of 1.0 M HCl (use a graduated cylinder) to this solution. This added acid will help to form large, filterable particles of precipitate. Heat this solution to near boiling (about 10 minutes on a hot plate). Do not heat to dryness.

5. While both solutions (steps 3 and 4) are still hot, slowly add the $BaCl_2$(aq) into the beaker of H_2SO_4(aq), with *vigorous stirring*. Add *slowly* so that large particles of precipitate will form. Rinse the stirring rod with your wash bottle before removing it from the beaker.

6. To further promote the formation of large particles of precipitate, you should heat the mixture in the beaker to near boiling on a hot plate and keep it at this temperature for 30 minutes. Do not heat to dryness. Stir occasionally, then allow it to cool slowly. This *digestion* procedure dissolves the smaller particles of solid, which later will reprecipitate on the remaining particles when the mixture cools.

7. Use the stirring rod to direct the liquid, and pour the mixture from the beaker onto the filter paper in the funnel. Wash the precipitate from the beaker with repeated 5 mL aliquots of distilled water. Make sure to also wash off the stirring rod so that all solid is quantitatively transferred. (If some of the fine precipitate passes through the filter paper, re-filter the filtrate through the same filter paper. Make sure to quantitatively transfer all of the filtrate to the funnel.) Carefully transfer the wet filter paper from the funnel onto a clean dry paper towel, and place it in your locker to dry until the next laboratory period. Be very careful with this transfer, since wet filter paper may tear. If a drying oven is available, place the wet filter on a large watch glass and dry in a 110°C oven for 1 to 2 hours.

8. Repeat steps 2 through 7 for a second determination.

9. Rinse all glassware that has been used for $BaCl_2$(aq) with small portions of H_2O. Discard these rinsings in the manner indicated for solutions in the disposal section.

10. During the next laboratory period, determine the mass of filter paper and $BaSO_4$(s) to the nearest mg. Subtract the recorded mass of the filter paper to determine the mass of the precipitate.

11. **DISPOSAL**

    Solid $BaSO_4$: Put precipitate and filter paper into a waste bottle labeled for solids.
    Filtrate and unused $BaCl_2$ solution: Add 6 M H_2SO_4 to precipitate all the barium as $BaSO_4$ and filter into a labeled waste bottle. The solid will be disposed in the same manner as that above, and the acid solution will be neutralized or disposed in a labeled waste bottle.

C. Volumetric Determination of Acid Concentration

1. Obtain a clean 50 mL buret. Rinse it twice with 5 mL portions of standardized 0.100 M NaOH solution. Make sure to run some of this rinse solution through the buret tip.

2. Fill the buret with 0.100 M NaOH. Do not fill the buret to the 0.00 mL mark. Instead, fill the buret nearly to this mark, making sure that the tip is filled, and read the liquid level (to one digit beyond the finest graduation; estimate the last digit).

3. Using a suction device and careful, safe procedure, pipet 10.00 mL of the **final solution** into a clean 250 mL Erlenmeyer flask. Add two or three drops of phenolphthalein indicator solution.

4. Titrate this sample of the final solution with 0.100 M NaOH(aq) until a *faint* pink color persists for over 15 seconds. Swirl the solution in the flask as you titrate. Wash down the walls of the flask with distilled water from your wash bottle to ensure that any splattered acid or base is returned to the titration mixture. A half drop of 0.100 M NaOH(aq) can be added by "hanging" a partial drop on the buret tip, touching this drop to the inside of the flask, and then washing the drop into the solution with distilled water. Record the volume (to ±0.01 mL) of the base used.

5. Repeat steps 2 through 4 with a second portion of the final solution, until the volume of titrant used in successive titrations agrees within 1%.

6. **DISPOSAL**

    Titration solutions: Since these are virtually neutral salt solutions, they may be flushed down the sink with running water.
    Unused H_2SO_4 and NaOH: Put in a labeled waste bottle, or mix unused acid and base in a large beaker. Add phenolphthalein and neutralize with 1 M H_2SO_4 (the first disappearance of pink color) or NaOH (first appearance of permanent pink color). Flush the resulting salt solution down the sink with running water.

D. Reduced Scale Volumetric Determination of Acid Concentration

As an alternative to the titration in Part C, a 10 mL buret may be used. This should be rinsed with small portions of 0.100 M NaOH. Pipet 2.00 mL of the final H_2SO_4 solution into a 25 mL volumetric flask. Titrate the solution by the procedure described in Part C.

Report   Name_____Section_____

Gravimetric and Volumetric Analysis

  DATA

B.   **Gravimetric Determination of Sulfate Ion**

	Trial 1	Trial 2	Trial 3	Trial 4
Mass of $BaSO_4$ and filter paper, g	_____	_____	_____	_____
Mass of filter paper, g	_____	_____	_____	_____
Mass of $BaSO_4$ precipitate, g	_____	_____	_____	_____
Millimoles of $BaSO_4$	_____	_____	_____	_____
Millimoles of H_2SO_4 that reacted	_____	_____	_____	_____
Volume of final solution reacted, mL	_____	_____	_____	_____
$[H_2SO_4]$, final solution, M	_____	_____	_____	_____
$[H_2SO_4]$, stock solution, M*	_____	_____	_____	_____
$[H_2SO_4]$, initial solution, M*	_____	_____	_____	_____
Average $[H_2SO_4]$, initial soln, M		_____		

C.   **Volumetric Determination of Acid Concentration**

	Trial 1	Trial 2	Trial 3	Trial 4
Molarity of NaOH(aq)		_____		
Initial NaOH buret reading, mL	_____	_____	_____	_____
Final NaOH buret reading, mL	_____	_____	_____	_____
Volume of NaOH(aq) used, mL	_____	_____	_____	_____
Millimoles of NaOH	_____	_____	_____	_____
Millimoles of H_2SO_4 titrated	_____	_____	_____	_____
Volume of final solution titrated, mL	_____	_____	_____	_____
$[H_2SO_4]$, final solution, M	_____	_____	_____	_____
$[H_2SO_4]$, stock solution, M*	_____	_____	_____	_____
$[H_2SO_4]$, initial solution, M*	_____	_____	_____	_____
Average $[H_2SO_4]$, initial soln, M		_____		

  * Calculated by taking the successive dilutions into account.

Sample Calculations

Report   Name_____Section_____

QUESTIONS

1. Calculate the concentrations of H_2SO_4 for the initial solution from your results for Parts B and C. Which result, B or C, is likely to be more precise? Explain.

2. There is an inherent flaw in the drying procedure for the gravimetric determination that causes the measured mass of $BaSO_4$ to be erroneously high. What is it and how could the procedure be improved?

3. If the volumetric procedure yields more accurate results, why perform the gravimetric procedure at all?

4. If the $BaSO_4$ residue was baked in an oven at 100°C for several hours rather than air drying, what effect would this have on its mass and how would the initial concentration of H_2SO_4 be affected (higher, lower, or no change)? Explain.

5. If a 20.00 mL pipet was unknowingly used in pipetting the final solution in Part C, how would the calculated initial concentration of H_2SO_4 be affected (higher, lower, or no change)? Explain.

Prelab   Name_____Section_____

Gravimetric and Volumetric Analysis

1. What is the principal purpose of this experiment?

2. Write a balanced equation for the reaction used in the *gravimetric* analysis part of this experiment.

3. What must be done with the solid barium sulfate waste?

4. 10.00 mL of approximately 6 M sulfuric acid is transferred to a 100-mL volumetric flask and diluted to the mark with distilled water and mixed. Then 10.00 mL of this solution is diluted further to 100 mL. What is the molarity of this last solution?

5. 10.00 mL of the final acid solution is reacted with excess barium chloride to produce a precipitate of barium sulfate (FW = 233.4 g/mol). The dry solid weighs 0.397 g. Use this mass and the dilution volumes to calculate the **actual** molarity of the sulfuric acid in the **initial** solution.

9 The Gas Laws

INTRODUCTION

The simple laws that govern the properties of gases can be readily demonstrated experimentally. In this experiment a Cenco gas laws apparatus (Figure 9-1) will be used to verify Boyle's law ($P \propto 1/V$) and Gay-Lussac's law ($P \propto T$). A knowledge of the uncertainties in the measuring instruments that are used will allow for the calculation of the percent uncertainty in the PV products and P/T ratios, which will be very nearly constant but not exactly so. This experiment also provides experience in the correct use of recorded data and the proper interpretation of such data.

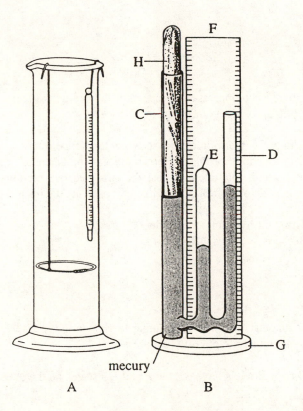

mecury

Figure 9-1. Cenco Gas Apparatus

THE APPARATUS

The equipment for this experiment (Figure 9-1) consists of a large hydrometer cylinder (A) and a Cenco gas apparatus (B). The Cenco gas apparatus consists of a three–tubed piece of glassware attached to a portion of a meter stick (F) and a metal base (G). The two outer tubes of the apparatus (C and D) are open to the atmosphere. The center tube (E) is closed at the top. A quantity of air (gas) is trapped in tube E between the mercury which fills the bottom of the apparatus and the sealed upper end of tube E. It is this air upon which measurements are made. A wooden plunger (H) is fitted into the larger of the two outside open tubes (C). Pushing down on this plunger increases the pressure on the air trapped in tube E and thus decreases its volume. Three different properties of the gas (air) are to be measured during this experiment. (**Caution:**

79

Tubes C and D should be stoppered when not in use and the mercury covered with a thin layer of glycerol to minimize exposure to mercury vapor.)

1. Temperature is measured by suspending a thermometer from the top of the shortened meter stick. (See your textbook for the definition of temperature.)

2. Pressure is the sum of (a) atmospheric pressure expressed in mmHg or torr (these are simply two different names for the same unit) and (b) the difference in the mercury levels between tube D and tube E. This difference will be in mmHg if the heights of the levels are in mm. Note that the top of tube E (the closed glass end) is at about 300 mm (or 30 cm). (See your textbook for the definition of pressure.)

3. Volume. The gas in tube E is cylindrical in shape. Tube E has a diameter of 0.300 cm and the air trapped therein has a length equal to the height of the top of the closed end (about 30 cm) minus the level of mercury in tube E. (Also record this level in cm. Thus a number is obtained ranging from about 5 to 30 cm.) The volume of a cylinder is given by $V = \pi r^2 h$, where $\pi = 3.14159$, r = radius of the tube = diameter/2, and h = the length of the gas cylinder.

SAFETY PRECAUTIONS

Review the safety rules on pages 1 and 2.

Be particularly careful in handling the Cenco apparatus, or alternative apparatus, so as to avoid breakage and spilling of mercury. If mercury is spilled, consult your instructor at once.

Be particularly careful when handling the thermometers used for this experiment. The breaking of a thermometer can occur easily and is a serious problem as mercury can be released into the environment. Mercury spills are one of the safety hazards that chemists fear the most because mercury vapor can quickly saturate the laboratory atmosphere. In addition, a broken thermometer presents the hazard of broken glass.

Use gloves when pouring hot water into the hydrometer cylinder.

PROCEDURE

A. Verification of Boyle's Law

Working at room temperature, manipulate the wooden plunger to change the pressure of the gas. Record the length of the air column in the inner tube for eight different pressures.

Make a plot of P versus $1/V$ to verify Boyle's law. Set the P and $1/V$ scales on the axes of your graph from your lowest to your highest values.

B. Verification of Gay-Lussac's Law

Pour hot (at least 80°C) water into the hydrometer cylinder. Immerse the Cenco apparatus into the cylinder and suspend a thermometer from the top of the measuring stick. Manipulate the plunger in such a way as to maintain a constant volume of gas. As the water cools, measure the pressure at eight different temperatures (about 6°C apart). To obtain the maximum number of (at least eight) measurements, start with the plunger all the way down when the hottest water is in the hydrometer cylinder.

To verify Gay-Lussac's law, make plots of the following:

1. P versus T(K) (Set T and P scales from your lowest to highest values.)

2. P versus T(K) (Set T and P scales from 0 to your highest value.)

Be sure to include all three plots in your laboratory report.

DISPOSAL

Avoid spilling any mercury or breaking a thermometer. If there is a mercury spill, notify your instructor. The spill will be cleaned up in a safe manner, preferably using a commercial spill kit.

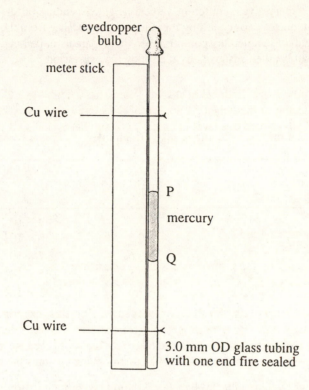

Figure 9-2. Alternative Apparatus for Boyle's Law Verification

ALTERNATIVE APPARATUS FOR VERIFICATION OF BOYLE'S LAW

An alternative to the apparatus shown in Figure 9-1 is constructed from a meter stick, a piece of 3.0 mm OD (0.180 mm ID) glass tubing, an eyedropper bulb to prevent mercury from spilling, and some copper wire, as shown in Figure 9-2. Remove the eyedropper bulb before making measurements. With this apparatus, the three variables are measured as detailed below.

1. Temperature is measured by hanging a thermometer from the meter stick so that the bulb is immersed in the water in the hydrometer cylinder. A large (500 mL or more) graduated cylinder may be used instead.

2. Volume is the volume of gas trapped in the sealed tube and is thus computed using the equation $V = \pi r^2 h$, where $\pi = 3.14159$, $r = 0.180$ cm/2, and $h =$ length from sealed end to mercury level in that tube (level Q).

3. Pressure is the sum of (a) the atmospheric pressure and (b) the vertical height, s, of the mercury column from level Q to level P. The vertical height should be measured in mmHg. The difference in pressure between the trapped gas and the atmosphere is constant as long as this apparatus is kept in a vertical position.

Variation of Pressure with Alternative Apparatus

The vertical height of the mercury column rather than its total length affects the pressure on the gas. Thus merely tilting the apparatus is sufficient to vary the pressure. s is the vertical height of the mercury column in the three diagrams in Figure 9-3. (The radius of the tube is exaggerated for clarity in the figure.)

The same amount of mercury gives a different vertical height depending on the angle of tilt. Measuring s directly can be difficult but one can compute s by measuring l, the distance from the bench top to the top of the glass tube. If H is the length of the mercury column when the apparatus is perpendicular to the bench top and L is the length of the glass tube, then s is given by the following expression.

$$s = l(H/L)$$

Sample results:

$$H = 112 \text{ mm}; L = 1226 \text{ mm}; P_{atm} = 763 \text{ mmHg}$$

l, mm	h, mm	s, mm	$P = P_{atm} + s$, mmHg	$P \times h$
1226	955	112	875	836×10^3
1067	972	97.5	860.5	836×10^3
849	996	77.6	840.6	837×10^3
622	1018	56.8	819.8	835×10^3
312	1050	28.5	791.5	831×10^3
Maximum deviation	9.4%		10.5%	0.7%

Since h is proportional to gas volume, the product $P \times h$ should be constant if Boyle's law is obeyed.

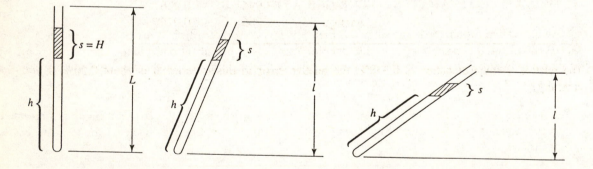

Figure 9-3. Variation of Pressure with Alternative Apparatus

CALCULATIONS

You should have made at least eight determinations for Part A and eight determinations for Part B.

1. For as many of these determinations as you have, write the volume (in cm^3), the pressure (in torr or mmHg), and the temperature (in K).

2. Calculate the value of the ratio PV/T for each determination.

3. Calculate the average value of the ratio *PV/T*.

4. Determine the deviation of each of the *PV/T* values from the average. (The deviation is the absolute value of the difference between the individual measurement and the average.)

5. Determine the average of the deviations.

6. Express this average of the deviations as a percentage of the average of the measurements. This percentage is the relative average deviation.

A sample follows.

	P, mmHg	*V*, cm³	*T*, K	*PV/T*	Deviation
1	936.1	1.349	296.2	4.263	0.003
2	905.1	1.399	296.2	4.275	0.009
3	886.1	1.427	296.2	4.269	0.003
4	836.1	1.505	296.2	4.248	0.018
5	823.1	1.526	296.2	4.241	0.025
6	816.1	1.554	296.2	4.282	0.016
7	807.1	1.561	296.2	4.253	0.013
8	976.1	1.561	259.2	4.242	0.024
9	963.1	1.561	353.2	4.257	0.009
10	938.1	1.561	341.2	4.292	0.026
11	923.6	1.561	335.2	4.301	0.035
12	903.1	1.561	329.2	4.282	0.016
13	883.1	1.561	323.2	4.265	0.001
14	868.1	1.561	317.2	4.272	0.006
15	848.1	1.561	311.2	4.254	0.012
			Total	63.996	0.236
			Average	4.266	0.014

The relative average deviation is 0.338%; the relative error in this experiment is about 0.50% if you work carefully.

Prelab   Name _____ Section _____

The Gas Laws

1.   What are the two potential safety hazards associated with the use of the thermometers?

2.   Given that the top of tube E is at the 28.5 cm mark on the meter stick and that the level of mercury in tube E is at the 17.4 cm mark, determine the volume of gas trapped in tube E. The diameter of tube of tube E is 0.300 cm.

3.   When the mercury level in tube D is higher than the mercury level in tube E, what is the relationship between the pressure of the gas trapped in tube E and the atmospheric pressure of the gas in the room?

4.   For the experimental conditions described in question 2, the mercury level in tube D is found to be at the 20.1 cm mark on the meter stick. Given that the atmospheric pressure in the room is 76.0 cm Hg, determine the pressure of the gas trapped in tube E.

5.   What three variables are measured in this experiment? In Part A, which of the three variables do you, the experimenter, manipulate, and which variable is held constant? In Part B, which of the three variables do you, the experimenter, manipulate, and which variable is held constant?

Report   Name _____ Section _____

Record your data with the correct number of significant figures and include units on each piece of data.

A. Verification of Boyle's Law

Room temperature = _____   Atmospheric pressure = _____

No.	Length of air column, cm	Volume of air, cm^3	Difference in height of Hg columns, mm	Pressure of gas, mmHg	$P \times V$
1	_____	_____	_____	_____	_____
2	_____	_____	_____	_____	_____
3	_____	_____	_____	_____	_____
4	_____	_____	_____	_____	_____
5	_____	_____	_____	_____	_____
6	_____	_____	_____	_____	_____
7	_____	_____	_____	_____	_____
8	_____	_____	_____	_____	_____

B. Verification of Gay-Lussac's Law

Atmospheric pressure = _____   Volume of gas = _____

No.	Difference in height of Hg columns, mm	Pressure of gas, mmHg	Temperature of gas, K	P/T
1	_____	_____	_____	_____
2	_____	_____	_____	_____
3	_____	_____	_____	_____
4	_____	_____	_____	_____
5	_____	_____	_____	_____
6	_____	_____	_____	_____
7	_____	_____	_____	_____
8	_____	_____	_____	_____

Report   Name _____ Section _____

	P, mmHg	V, cm^3	T, K	PV/T	Deviation
1	_____	_____	_____	_____	_____
2	_____	_____	_____	_____	_____
3	_____	_____	_____	_____	_____
4	_____	_____	_____	_____	_____
5	_____	_____	_____	_____	_____
6	_____	_____	_____	_____	_____
7	_____	_____	_____	_____	_____
8	_____	_____	_____	_____	_____
9	_____	_____	_____	_____	_____
10	_____	_____	_____	_____	_____
11	_____	_____	_____	_____	_____
12	_____	_____	_____	_____	_____
13	_____	_____	_____	_____	_____
14	_____	_____	_____	_____	_____
15	_____	_____	_____	_____	_____
16	_____	_____	_____	_____	_____

Total _____ _____

Average _____ _____

Relative average deviation _____%

The relative error in this experiment is about 0.50%.

Sample Calculations

port   Name _____ Section _____

QUESTIONS

1. Based on the values of $P \times V$ determined in Part A of the procedure and the plot of P versus $1/V$, what is the relationship between P and V? Has Boyle's law been verified?

2. Based on the values of P/T determined in Part B of the procedure and the plot of P versus T, what is the relationship between P and T? Has Gay-Lussac's law been verified?

3. Determine the value of absolute zero from the x-intercept for the plot of P versus T where both axes begin at zero? What is this temperature in °C? How does this value compare to the known value of absolute zero? What is the relative error?

4. What is the average deviation of the ratio PV/T? The relative error, which is a measure of accuracy, in this experiment is about 0.50%. Are your results more precise or more accurate?

5. Use your average value of PV/T and the ideal gas constant, R (0.08205 L atm mol^{-1} K^{-1}), to determine the number of moles of gas present in tube E.

10 Evaluation of the Gas Law Constant

INTRODUCTION

The presentation of the ideal gas law $PV = nRT$ can be found in your textbook. This law will be used in this experiment to evaluate R, the gas law constant. If R is to be determined, the other parameters P, V, n and T in the equation must be available from the experiment–that is, it must be possible to measure them in the laboratory.

The procedure of this experiment is based on the chemical reaction between magnesium metal and hydrochloric acid to produce hydrogen gas. The volume, pressure, and temperature under which the hydrogen is collected will be measured. From the known quantity of magnesium used and the stoichiometry of the reaction the number of moles of hydrogen produced can be calculated.

$$Mg(s) + 2\ HCl(aq) \rightarrow MgCl_2(aq) + H_2(g)$$

Since the hydrogen is collected in a eudiometer tube over an aqueous solution (see following procedure), the gas pressure in the tube after the reaction has ceased is the sum of the hydrogen gas pressure and the vapor pressure of water (Dalton's Law). In order to obtain the pressure of the hydrogen gas, the vapor pressure of water, P_{H2O}, at the temperature of the measurement must be subtracted from the atmospheric pressure, P_{atm}. Thus the pressure of the hydrogen is given by

$$P_{H2} = P_{atm} - P_{H2O}$$

In case the liquid levels (step 3) cannot be equalized after the reaction has ceased, a further correction will be required since the pressure of the gases in the tube (hydrogen and water vapor) will then not be equal to the atmospheric pressure. If this is the case, the difference in levels must be measured with a meter stick as accurately as possible. Note that the graduations on the tube are in milliliters, not millimeters. *You must use a meter stick.* This difference, which represents the desired pressure difference, must be converted to mmHg. This can be accomplished by dividing the measured level difference in millimeters by 13.5 (the ratio of the densities of Hg and the aqueous solution). This difference must then be subtracted from the atmospheric pressure. Thus, if the levels cannot be equalized, the pressure of hydrogen must be obtained from the following expression.

$$P_{H2} = P_{atm} - P_{H2O} - P_{level\ difference}$$

where $P_{level\ difference} = \dfrac{\text{Difference in heights in mm as measured}}{13.5}$

Because several measurements are readily obtained within a 2- to 3-hour time period, the precision of this experiment may be readily determined. In addition, the accuracy can be determined since the accepted value of R is known. Proper evaluation of the data recorded for this experiment is an important and meaningful part of the experiment's objectives of verifying the gas law constant and demonstrating a means of measuring it simply in the laboratory.

Vapor Pressure (P) of Water at Various Temperatures

t, °C	P, mmHg	t, °C	P, mmHg	t, °C	P, mmHg	t, °C	P, mmHg
13.0	11.2	19.0	16.5	25.0	23.8	31.0	33.7
14.0	12.0	20.0	17.5	26.0	25.2	32.0	35.7
15.0	12.8	21.0	18.7	27.0	26.7	33.0	37.7
16.0	13.6	22.0	19.8	28.0	28.3	34.0	39.9
17.0	14.5	23.0	21.1	29.0	30.0	35.0	42.2
18.0	15.5	24.0	22.4	30.0	31.8	36.0	44.6

SAFETY PRECAUTIONS

Review the safety rules on pages 1 and 2.

Hydrogen is highly flammable. There should be no open flames or electrical sparks in the area.

Be very careful when adding the concentrated HCl to the eudiometer. Avoid skin contact with HCl and use concentrated HCl only in a **fume hood**. Rinse any spills thoroughly.

Needless repetition and accidents can be avoided if the Mg ribbon is wrapped carefully with Cu wire so that it doesn't break free while it is reacting.

PROCEDURE

1. Calculate the mass of magnesium necessary to evolve 80 mL of H_2 at STP. Then weigh approximately this quantity of Mg ribbon on the top-loading balance to the nearest mg (±0.001 g).

2. In the hood add approximately 8 mL of concentrated HCl to a eudiometer tube. With your wash bottle, wash down into the tube any acid that might have adhered to the eudiometer walls. Coil the weighed strip of Mg ribbon into the eudiometer tube, approximately 5 to 10 cm from the open end. (Copper wire wrapped around the ribbon helps to keep it in place.) Fill the tube with water and invert into a suitable vessel nearly filled with water. The eudiometer volume should read zero if the tube is filled completely with water. Record the initial reading in your laboratory notebook. Clamp the tube in position (Figure 10-1).

3. The concentrated HCl will gradually diffuse down to the Mg and react with it. The Mg may break free from the copper wire. Tilt the tube to ensure that the Mg does not stick to the sides of the tube. After reaction has ceased, if possible, adjust the liquid level in the tube to the level of liquid in the vessel. The gas pressure in the tube is then equal to atmospheric pressure. If the difference in levels is large, it must be measured as shown in Figure 10-1.

4. Record the temperature of the water in the vessel at 1-min intervals for 3 min. Take the average value as your gas temperature.

5. Repeat the experiment for a total of four determinations.

6. Calculate the number of moles of magnesium used as a reactant.

$$\text{Number of moles of Mg} = \frac{\text{Mass of Mg(g)}}{24.31 \text{ g/mol}}$$

7. Calculate the corrected pressure for the experiment as described above using the observed barometric pressure for the day.

11 Thermochemistry: The Heat of Reaction

INTRODUCTION

Chemical thermodynamics deals with the energy changes that accompany chemical reactions. Such energy changes are among the factors that determine the following:

1. How fast a chemical reaction takes place–that is, the problem of chemical kinetics (see Experiment 15).
2. How complete the reaction will be–that is, the position of chemical equilibrium (see Experiment 16).

Thermochemistry concerns the energy changes that are manifested as the enthalpy change of reaction, ΔH. ΔH is the heat given off by the reaction at constant pressure. A reaction in which heat is lost by the reactants to the surroundings has a negative ΔH and is said to be exothermic; one in which heat is absorbed has a positive ΔH and is endothermic.

The general term, *enthalpy of reaction*, may be classified into more specific categories:

1. The enthalpy of formation is the quantity of heat involved in the formation of 1 mole of the substance in its standard state directly from its constituent elements in their standard states.

2. The enthalpy of combustion is the quantity of heat evolved per mole of a combustible substance, such as carbon or methane, undergoing a reaction with excess oxygen.

3. The enthalpies of solution, vaporization, fusion, and sublimation are concerned with changes in state or solvation of molecules or ions.

4. The enthalpy of neutralization is the heat evolved when 1 mole of water is produced by the reaction of an acid and base.

In this experiment, we shall measure the enthalpies of neutralization of HCl and of $HC_2H_3O_2$ solutions with NaOH solution, the enthalpy of solution of NaOH(s), and the enthalpy of neutralization of NaOH(s) with HCl solution. Comparison of calculated results for different parts of the experiment should permit you to verify the generalization known as Hess's law of constant heat summation, which states that a process proceeds through one or several steps, the enthalpy change for the overall process can be calculated by summing the enthalpy changes for the individual steps.

Heat measurements are performed by carrying out the reaction in a calorimeter (Figure 11-1), in which the temperature change and mass of solution are measured. The purpose of the calorimeter is to prevent the gain of heat from the surroundings or the loss of heat to the surroundings. As a chemical reaction proceeds, heat is given off by the reaction and this same heat is gained by the solution. Thus the temperature of the solution increases. If we assume that the heat gained or lost by the calorimeter is insignificant, the heat of the reaction, q_{rxn}, is the negative of the heat that is gained by the solution, q_{soln}. The heat gained by the solution is calculated from the equation: $q_{soln} = mC\Delta t$, where m and Δt are the mass and temperature change of the solution. The specific heat, C, is the quantity of heat required to raise the temperature of one gram of material (in this case, the material is the solution) by one degree Celsius (J g^{-1} °C^{-1}). The specific heat is given for the different parts of this experiment. The heat given off by the reaction and gained by the solution is measured in units of joules. The change in enthalpy, ΔH, of a reaction is the heat evolved per mole of reactant. The ΔH is determined by dividing the q_{rxn} by the number of moles of reactant or product. In this experiment, the number of moles of water produced in Parts B and C equals the number of moles of each reactant.

The assumption that no heat is gained by the calorimeter when the reaction occurs is not strictly true.   However, not including the heat gained by the calorimeter introduces a typical error of 2% or less.   This amount of error is well within the experimental error associated with the measurement of the temperature change in the experiment.

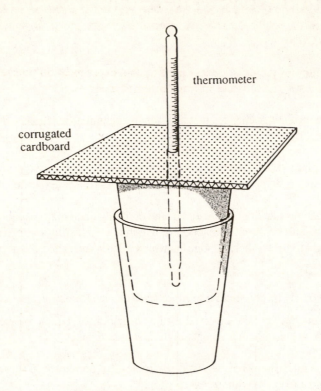

Figure 11-1. Simple Calorimeter for Measuring Heat of Reaction

DETERMINATION OF TEMPERATURE

Three factors make it difficult to determine temperatures quickly in this experiment.

1. The calorimeter is not a perfect insulator, and heat "leaks" out of it (causing the temperature to drop by about 0.1°C/min).

2. It takes at least 1 min for the calorimeter to reach the same temperature as the mixture within it.

3. In Parts D and E, it takes about 3 min for the NaOH(s) to dissolve.

All three effects are evident in this set of student data from Part E of the experiment.

Time, min	1	2	3	4	5	6	7	8	9	10	11	12
Temp, °C	16.5	16.6	16.7	mix	22.0	24.0	26.4	28.5	28.0	28.0	27.8	27.7

These data are graphed in Figure 11-2. Notice at times 1, 2, and 3 that the temperature is slowly rising because the temperature of the HCl is cooler than that of the laboratory. To determine the temperature of the HCl at the time of mixing, we extrapolate these points forward in time (draw the best straight line through them) and obtain 16.8°C at the time of mixing. From time 5 to time 8 the NaOH is dissolving and reacting with the HCl and the temperature rises rapidly to 28.5°C. It then drops back quickly to 28.0°C, probably because it took this last minute to heat the calorimeter. Now the temperature begins to fall slowly and regularly (about 0.1°C/min), and we can extrapolate through these data backward in time to find out what the temperature at the time of mixing would have been if the NaOH had dissolved instantaneously and the calorimeter had warmed instantaneously. Observe carefully in Figure 11-2 that data are collected for 8 min after mixing in order to obtain good results. Do not stop collecting data too soon. In fact, it is a good idea to plot your data as you collect them. You should make one graph like the one in Figure 11-2 for each set of data you collect.

SAFETY PRECAUTIONS

Review the safety rules on pages 1 and 2.

Be particularly careful when handling the sensitive thermometers used for this experiment. The breaking of a thermometer can occur easily and is a serious problem as mercury can be released into the environment. Mercury spills are one of the safety hazards that chemists fear the most. In addition, a broken thermometer presents the hazard of broken glass.

If you spill any of the solutions (acids or bases) used in this experiment on yourself or your work area, wash thoroughly with running water and tell your instructor immediately.

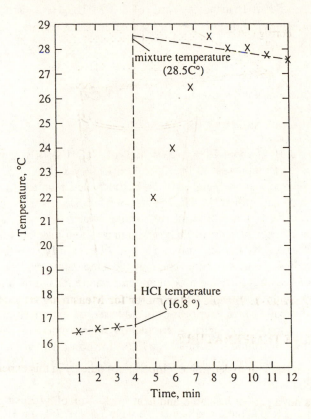

Figure 11-2. Temperature-Time Plot

PROCEDURE

A. Thermometer Calibration

Prepare two calorimeters, each similar to the one illustrated in Figure 11-1, as directed. Compare your two thermometers by immersing them together in water at room temperature for 1 min, and reading the temperature of each as nearly as possible to the nearest 0.1°C. Be careful to avoid parallax in your readings. Always use the same thermometer in the calorimeter in which the temperature change occurs, and in all subsequent readings apply any necessary correction to the other, so that the readings of both thermometers will correspond.

B. The Enthalpy of Neutralization of HCl(aq) and NaOH(aq)

Place 50.0 mL of 1.0 M HCl in one calorimeter and 50.0 mL of 1.0 M NaOH in the other calorimeter. With the lids and thermometers in place, read the temperatures (±0.1°C) for 3 min at 1-min intervals; quickly mix the NaOH thoroughly into the HCl solution, and continue the readings for 4 min at 1-min intervals.

Extrapolate the temperatures to the time of mixing for each solution as in Part A, and calculate the enthalpy of neutralization per mole of water produced. (The density of the 0.5 M NaCl produced is 1.02 g/mL, and its specific heat is 4.00 J g^{-1} °C^{-1}.)

DISPOSAL

Solutions: Add phenolphthalein and neutralize with 1 M HCl (the first disappearance of pink color) or NaOH (first appearance of permanent pink color). Flush the resulting salt solution down the sink with running water.

C. The Enthalpy of Neutralization of HC$_2$H$_3$O$_2$ (aq) and NaOH(aq)

Repeat the procedure of Part B using 50.0 mL of 1.0 M HC$_2$H$_3$O$_2$ and 50.0 mL of 1.0 M NaOH. Calculate the enthalpy of neutralization as before. (Assume the same density and specific heat as for NaCl in Part B.)

DISPOSAL

Solutions: Follow instructions for disposal in Part B.

D. The Enthalpy of Solution of NaOH(s)

Carefully weigh (to ±0.01 g) about 2.00 g (0.05 mol) of NaOH(s). [Because of its hygroscopic nature, weigh this by difference, in a stoppered 50 or 125 mL Erlenmeyer flask used as a weighing bottle. Your instructor will tell you the approximate number of NaOH(s) pellets required to assist in estimating the mass needed. Be sure to clean up any spilled NaOH(s). This solid absorbs water from the air and forms a slippery solution that is also corrosive.] Place 50.0 mL of distilled water in your calorimeter. With the lid and thermometer in place, read the temperature (0.1°C) for 3 min at 1-min intervals; then add the NaOH(s), replace the lid and thermometer, and gently swirl the mixture and stir it **carefully** with the thermometer to dissolve the NaOH as quickly as possible. At the same time continue the temperature readings at 1-min intervals for at least 9 min total (at least three readings after the maximum temperature reading is attained). Because of the time required for solution and complete mixing, the proper estimate of the temperature for complete solution at the time of mixing is more difficult. Make your best estimate based on extrapolations of temperatures before and after mixing as explained in the section on the determination of temperature on page 96. A plot of your data is essential. Calculate the heat of solution per mole NaOH(s) to form a 1.0 M NaOH solution. (Note that you have about 52 g of solution. The specific heat of 1.0 M NaOH is 3.90 J g^{-1} deg^{-1}.)

DISPOSAL

NaOH solution: Add phenolphthalein and neutralize with 6 M H$_2$SO$_4$ or HCl (the first disappearance of pink color). Flush the resulting salt solution down the sink with running water.

E. The Enthalpy of Reaction of HCl(aq) and NaOH(s)

Carefully weigh about 2.00 g (to 0.01 g) or slightly less of NaOH(s). Use a stoppered Erlenmeyer flask as in Part D. Measure about 55 mL of 1.0 M HCl (which provides a small excess of HCl to react with all the NaOH used) in a 100 mL graduated cylinder, and dilute this to 100.0 mL, as precisely as you can. Transfer this completely to your calorimeter and, with the lid and thermometer in place, take temperature readings once each minute for 3 min. Then add the NaOH(s), replace the lid and thermometer, gently swirl the mixture and stir it carefully with the thermometer to dissolve the NaOH as quickly as possible, and at the same time continue temperature readings for 9 min total at 1-min intervals (at least three readings after the maximum temperature reading is attained). Calculate the enthalpy of the reaction per mole of water formed. (Note that the density and the specific heat of the NaCl(aq) are about the same as in Part B. Calculate on the basis of the mass of NaOH(s) used, as a slight excess of HCl was specified to ensure complete neutralization. You may check this point with phenolphthalein.)

DISPOSAL

Solutions: Follow instructions for disposal in Part B.

Example Calculation

A 60.0 mL solution of 1.0 M HNO_3(aq) and 60.0 mL of 1.0 M NaOH(aq) are mixed in a calorimeter. The resulting change in temperature is measured to be 6.8°C. Determine the change in enthalpy for this reaction. The specific heat and density of the resulting solution are 4.0 J g^{-1}°C^{-1} and 1.03 g/mL respectively.

Mass of solution, m = 120.0 ml × 1.03 g/mL = 124 g

Heat gained by solution, q_{soln} = $mC\Delta t$ = (124 g) × (4.0 J g^{-1}°C^{-1}) × (6.8°C) = 3.4 × 10^3 J

Heat of the reaction, q_{rxn} = $- q_{soln}$ = $- 3.4 × 10^3$ J

Moles of HNO_3 = moles of NaOH = moles of H_2O = 0.0600 L × 1.0 M = 0.060 moles

Change in enthalpy, ΔH = ($- 3.4 × 10^3$ J)/(0.060 moles) = $- 5.7 × 10^4$ J/mol = $- 57$ kJ/mol

Report Name _____ Section _____

Lab Partner _____

B. The Enthalpy of Neutralization of HCl(aq) and NaOH(aq)

Place 50.0 mL each of 1.0 M HCl and 1.0 M NaOH at room temperature in respective calorimeters; temperature readings are as shown below.

	Trial 1 Calorimeters		Trial 2 Calorimeters		Trial 3 Calorimeters	
	HCl	NaOH	HCl	NaOH	HCl	NaOH
1.	_____	_____	_____	_____	_____	_____
2.	_____	_____	_____	_____	_____	_____
3.	_____	_____	_____	_____	_____	_____
4.	mix		mix		mix	
5.	_____		_____		_____	
6.	_____		_____		_____	
7.	_____		_____		_____	
8.	_____		_____		_____	

Temperatures at Time of Mixing
(from Graph or by Extrapolation)

HCl _____ _____ _____

NaOH _____ _____ _____

Mixture _____ _____ _____

	Trial 1	Trial 2	Trial 3
Heat gained by mixed solutions*	_____ J	_____ J	_____ J
Heat of the reaction (negative of heat gained)	_____ J	_____ J	_____ J
Enthalpy of neutralization (per mole of water formed)	_____ J/mol	_____ J/mol	_____ J/mol

*For temperature gain, $\Delta t°C$, use the temperature of the mixture minus the average temperature of HCl and NaOH. The density of 0.5 M NaCl produced is 1.02 g/mL, and its specific heat is 4.00 J g^{-1} $°C^{-1}$.

Sample Calculations

Report

Name _____ Section _____

Lab Partner _____

C. The Enthalpy of Neutralization of $HC_2H_3O_2$ (aq) and NaOH(aq)

Place 50.0 mL each of 1.0 M $HC_2H_3O_2$ and 1.0 M NaOH at room temperature in respective calorimeters; temperature readings are as shown below.

	Trial 1 Calorimeters		Trial 2 Calorimeters		Trial 3 Calorimeters	
	$HC_2H_3O_2$	NaOH	$HC_2H_3O_2$	NaOH	$HC_2H_3O_2$	NaOH
1.	_____	_____	_____	_____	_____	_____
2.	_____	_____	_____	_____	_____	_____
3.	_____	_____	_____	_____	_____	_____
4.	Mix		Mix		Mix	
5.	_____		_____		_____	
6.	_____		_____		_____	
7.	_____		_____		_____	
8.	_____		_____		_____	

Temperatures at Time of Mixing
(from Graph or by Extrapolation)

$HC_2H_3O_2$ _____ _____ _____

NaOH _____ _____ _____

Mixture _____ _____ _____

	Trial 1	Trial 2	Trial 3
Heat gained by mixed solutions*	_____ J	_____ J	_____ J
Heat of the reaction (negative of heat gained)	_____ J	_____ J	_____ J
Enthalpy of neutralization (per mole of water formed)	_____ J/mol	_____ J/mol	_____ J/mol

*The $NaC_2H_3O_2$ solution has about the same density and specific heat as the 0.5 M NaCl, Part B.

Sample Calculations

Report   Name _____ Section _____

Lab Partner _____

D. The Enthalpy of Solution of NaOH(s)

Record the temperature readings for 50.0 mL of water placed in a calorimeter, before and after adding NaOH(s), as shown below. Record the mass of NaOH(s) in the lower table.

Trial 1 Calorimeter Water		Trial 2 Calorimeter Water	
1. _____	7. _____	1. _____	7. _____
2. _____	8. _____	2. _____	8. _____
3. _____	9. _____	3. _____	9. _____
4. Mix	10. _____	4. Mix	10. _____
5. _____	11. _____	5. _____	11. _____
6. _____	12. _____	6. _____	12. _____

Temperatures at Time of Mixing
(from Graph or by Extrapolation)

Water   _____   _____   _____

Mixture   _____   _____   _____

	Trial 1	Trial 2
Mass of NaOH(s) + flask	_____ g	_____ g
Mass of flask	_____ g	_____ g
Mass of NaOH(s)	_____ g	_____ g
Heat gained by solution*	_____ J	_____ J
Heat of reaction (negative of heat gained)	_____ J	_____ J
Enthalpy of solution [per mole of NaOH(s)]	_____ J/mol	_____ J/mol

*The mass of solution is about 52 g (H_2O + NaOH). The specific heat of the NaOH solution is about 3.90 J g^{-1} °C^{-1}.

Sample Calculations

12 Spectrophotometric Analysis of Commercial Aspirin

INTRODUCTION

The interactions of electromagnetic radiation with matter in various regions of the spectrum provide us with an extremely valuable tool for both qualitative and quantitative analysis as well as for revealing molecular structure. From x-ray diffraction to ultraviolet to infrared spectroscopy, we have available to us techniques to aid in the study of problems of analysis, synthesis, or structure that were simply not available prior to this century. The development of such spectroscopic techniques during the past several decades has thus provided a tremendous stimulus to chemical research and to our overall knowledge.

In this experiment we shall use visible electromagnetic radiation, or white light, as a means to analyze the percent composition of commercial grade aspirin. Visible light and other forms of electromagnetic radiation (x-ray, ultraviolet, infrared, etc.) obey the Planck relationship.

$$E = h\nu \tag{12.1}$$

All such forms of light may be considered to be waves of electromagnetic energy propagated through space (Figure 12-1). Such waves may be described in terms of their wavelength, frequency, and speed. The wavelength, λ, may be defined as the distance between crests of the wave (Figure 12-1), and the frequency, ν, is the number of crests passing a given point in space each second. If the wavelength is specified in m and the frequency in cycles per sec (s^{-1}), the product of λ and ν has the units of m/s and is the speed of light, 3.00×10^8 m/s. [Note: $1\ s^{-1} = 1$ Hertz (Hz).]

$$\lambda\nu = c \tag{12.2}$$

In the visible region, wavelength is usually measured in terms of angstroms ($1\ \text{Å} = 10^{-10}$ m) or nanometers ($1\ \text{nm} = 10^{-9}$ m). The visible region usually is taken to be the region from 380 to 750 nm. Within this region light is composed of the various spectral colors (Table 12-1). White light, or sunlight, is composed of all these wavelengths and is known as *polychromatic light*. Light of a single wavelength is known as *monochromatic light*.

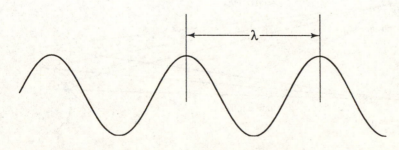

Figure 12-1. Wavelength

Table 12-1. Correlation Between Wavelength, Color, and Complementary Color in the Visible Region

Wavelength, nm	Color	Complementary Color
380–435	Violet	Yellow-green
435–480	Blue	Yellow
480–490	Green-blue	Orange
490–500	Blue-green	Red
500–560	Green	Purple
560–580	Yellow-green	Violet
580–595	Yellow	Blue
595–650	Orange	Green-blue
610–750	Red	Blue-green

When white light comes into contact with an object, it may be reflected by, absorbed by, or transmitted through the object. If a portion (certain wavelengths) of the incident radiation is absorbed by the object, the transmitted or reflected light will appear as the complementary color of those wavelengths, as shown in Table 12-1.

Since many substances form colored solutions, a comparison of the intensity of the color of a solution of known concentration with that of a solution of unknown concentration offers a convenient way for the quantitative estimation of the concentration of the colored substance in the unknown solution. If the comparison is done visually, the method is called *colorimetry*. If a photoelectric cell is used in place of the human eye, the method is called *photometry*. In the technique used in this experiment, when monochromatic light is used as the source and a photoelectric cell is used as the detector, the method is known as *spectrophotometry*. Figure 12-2 shows the optics of a typical spectrophotometer, such as the Spectronic 20. The white light emanating from the tungsten lamp passes through the entrance slit and is reflected by a diffraction grating. The grating is a dispersing element, acting like a prism to separate the white light into its component wavelengths. In setting the wavelength, the grating is rotated by means of a cam, so that the desired wavelength passes through the exit slit. The monochromatic light that passes through the exit slit goes on through the sample and finally strikes the measuring photoelectric cell, where the light energy is converted to an electric signal.

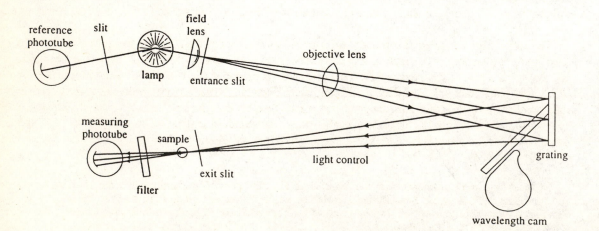

Figure 12-2. Optics of a Spectrophotometer

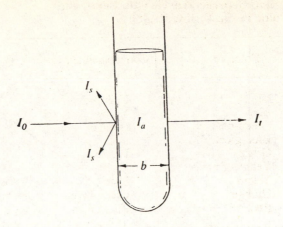

Figure 12-3. The Absorption, Scattering, and Transmission of Light by a Solution

When light of intensity I_0 impinges on a solution contained in a transparent cell, some of the light is scattered, or reflected, by the walls of the container and by the particles in solution, I_s; some is absorbed, I_a; and the rest is transmitted through the solution, I_t. Figure 12-3 shows the distribution of the incident light passing through a cell of thickness b where $I_0 = I_s + I_a + I_t$.

Under most experimental conditions, I_s is small and can be kept constant by using matched cells and by always positioning the cells in the instrument in the same way. Thus $I_0 - I_s$ is approximated as I_0. The absorbed light, I_a, is related to the concentration of the absorbing material in the solution. Thus, by measuring I_0 and I_t, the value of I_a can be calculated, and the concentration of the absorbing substance can be determined. I_a cannot be measured directly, but it can be related to the percent transmittance, %T, defined as $I_t/I_0 \times 100$, which is measured by the phototube. The absorbance A, defined as $\log I_0/I_t$, is proportional to the concentration, c, of the absorbing substance in the solution. The *Beer-Lambert law* provides the mathematical correlation between absorbance and concentration. It is usually stated as

$$A = abc \qquad [12.3]$$

In equation [12.3], a and b are constants that allow the proportionality $A \propto c$ to be converted to an equation. The constant a is known as the *absorptivity* and is characteristic of the absorbing solute. When the concentration is in molarity units, it is called *molar absorptivity*. The pathlength of the radiation through the cell is identical with the cell thickness b (Figure 12-3). Since cells of constant thickness are used, the absorbance A will depend linearly on the concentration of the absorbing solute for which a will be constant.

The relationship between absorbance and concentration serves as the basis for the quantitative analysis of a great many substances. In practice, the absorbances of a series of solutions of known concentrations are measured and a plot of absorbance versus concentration is prepared (Figure 12-4). Such a plot is known as a Beer's law graph and also represents a calibration curve for the particular system being studied. An unknown solution containing the same absorbing substance may then be analyzed by measuring its absorbance, locating its A value on the calibration curve, and reading the corresponding concentration. In Figure 12-4, the dotted line shows how the curve is used to find the concentration of an unknown solution with an absorbance of 0.82 ($c = 4.3 \times 10^{-4}$ M).

In recording measurements with the Spectronic 20, the linear % transmittance scale is more conveniently and precisely read than the logarithmic absorbance scale. However, since it is the absorbance that is linearly dependent on concentration, %T readings must be converted to the corresponding A values. These can be readily calculated from the formula

$$A = 2 - \log(\%T) \qquad [12.4]$$

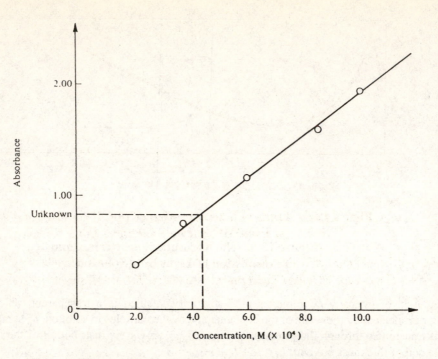

Figure 12-4. A Typical Beer's Law Plot

DETERMINATION OF ACETYLSALICYLIC ACID

Acetylsalicylic acid, commonly known as aspirin, is hydrolyzed rapidly in basic medium to yield salicylate dianion, as shown here:

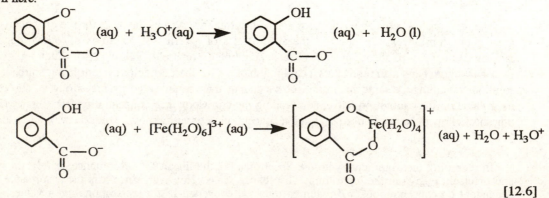

$$[12.5]$$

Acidification, followed by addition of iron(III) ion, yields the soluble tetraaquasalicylatoiron(III) complex ion shown here:

$$[12.6]$$

This intensely violet solution exhibits the transmittance spectrum shown in Figure 12-5. The complex in solution has a very high molar absorptivity, which allows it to be used as a sensitive indicator for the presence of salicylate species. The composition of this complex is sensitive to pH, and the solutions must be maintained in the pH range 0.5 to 2.0 to avoid formation of the di- and trisalicylato complexes of iron(III).

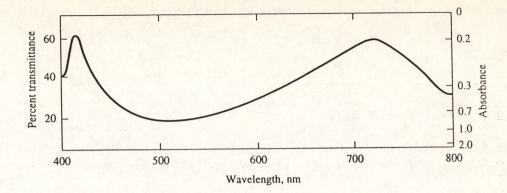

**Figure 12-5. Transmittance/Absorbance Spectrum of a Solution
of Iron(III)-Salicylate Complex Ion**

The wavelength of minimum % T just above 500 nm corresponds to the λ where a is at its maximum value and where A versus C will have its widest linear range. In addition, maximum sensitivity is realized at this λ.

Calculations

By utilizing the assumptions implicit in the discussion–that is, quantitative hydrolysis of the acetylsalicylic acid and a ratio of one-to-one salicylate ion to iron(III) ion in the complex (equation [12.6])–the concentration of the complex in each reference solution can be calculated.

As an example, assume 0.400 g of acetylsalicylic acid (aspirin) is treated as outlined in the procedure. The concentration of complex in the stock solution can be found as follows.

Molar mass of acetylsalicylic acid ($CH_3COOC_6H_4COOH$) = 180.2 g/mol.

$$0.400 \text{ g aspirin} \times \frac{1 \text{ mol}}{180.2\text{g}} = 2.22 \times 10^{-3} \text{ mol}$$

Upon hydrolysis, acidification, and dilution to 100 mL, the molarity of the solution is

$$M = 2.22 \times 10^{-3} \text{ mol} \times \frac{1000 \text{ mL/L}}{100 \text{ mL}}$$

$$= 2.22 \times 10^{-2} \text{ M}$$

This stock solution is diluted in varying proportions to yield the standard solutions A, B, C, D, and E. The concentration of complex in solution A can be found through the relation

$$M_1 V_1 = M_2 V_2$$

where M_1 and M_2 are the molarities of the respective stock and diluted solution, and V_1 and V_2 are the respective volumes of each. For the example above,

$$(0.500 \text{ mL})(2.22 \times 10^{-2} \text{ M}) = (10.0 \text{ mL})(x \text{ M}),$$

$$x = 1.11 \times 10^{-3} \text{ M}$$

This calculation is repeated for each dilution of the stock solution.

To determine the quantity of aspirin in the tablets, the absorbance of its solution, as the complex, is determined and the concentration of complex in the solution is read from the Beer's law plot. Knowing the sequence of dilutions followed, we can work backward through the calculations to obtain the mass of pure aspirin in the tablet.

SAFETY PRECAUTIONS

Review and follow the safety rules on pages 1 and 2.

Eye protection must be worn at all times.

Be particularly careful when heating the aspirin-NaOH mixture. NaOH is caustic and corrosive. This solution has a tendency to spatter when heated. If any spills occur, rinse and wipe thoroughly.

PROCEDURE

A. Operating Instructions for the Spectronic 20

Instrument Start Up. Turn on the instrument by rotating the left-hand knob, the amplifier control, clockwise. On some models of instruments, a red jewel will light at this point. Allow about 10 to 15 min for the instrument to warm up before recording any measurements.

Setting 0% Transmittance. After the instrument has warmed up, adjust the left-hand knob so that the meter needle reads 0 on the percent transmittance scale. This adjustment is made with no cell in the holder, so that no light strikes the phototube.

Cleaning Cuvettes and Obtaining a Matched Pair. While the instrument is warming up, rinse two test tube cells, called *cuvettes*, with distilled water three times. Fill each about two-thirds full with distilled water and insert one cuvette into the sample holder. Adjust the right-hand knob so that the needle reads about 50% transmittance. Replace this cuvette with the other. If the %*T* reading of the second cuvette is not within 1% of the value read for the first cuvette, repeat using other cuvette combinations until the %*T* readings of two cuvettes fall within 1% of each other. Do not handle the lower portion of the cuvettes because smudges or droplets of solution will change I_s. Wipe off the outside of the cuvettes with laboratory tissue.

Setting 100% Transmittance. Adjust the wavelength to the desired setting by using the knob on the top of the instrument. Turn the right-hand knob, the light control, counterclockwise almost to its limit before inserting a cuvette into the sample holder. Insert the cuvette containing the blank into the holder, matching up the index line on the cuvette exactly with the index line on the holder. Close the top of the holder. Adjust the right–hand knob clockwise until the needle reads 100 on the percent transmission scale. Immediately remove the test tube to avoid fatiguing the phototube.

Check of 0% T and 100% T. After the cell is removed from the holder, an occluder automatically drops into the light beam. The needle should then read 0. Each time the wavelength is changed and during all measurements at the same wavelength, establish or check the 0% *T* and 100% *T* settings.

Sample Measurements. Insert the cuvette containing the sample into the holder, matching the index line on cuvette and on the cell holder. Immediately record the percent transmittance to three significant figures. Remove the cuvette and note whether the 0 setting has changed. Repeat the measurements, and if there is a small difference between the readings, average the readings. After completing all measurements, rinse each cuvette with distilled water.

B. Preparation of a Standard Beer's Law Curve

On weighing paper, weigh to the nearest mg (±0.001 g) approximately 0.16 g of reagent grade acetylsalicylic acid. Record the actual mass of acetylsalicylic acid in your notebook. Transfer the sample quantitatively to a 125 mL Erlenmeyer flask. Add 5 mL of 1 M sodium hydroxide and heat the mixture to boiling. Care should be exercised to avoid spattering and loss of contents. (CAUTION! NaOH is harmful to skin and eyes.) If necessary, the inside walls can be rinsed down with distilled water to ensure quantitative hydrolysis of the acetylsalicylic acid. Allow this solution to cool.

Quantitatively transfer the resulting solution of sodium salicylate to a 100 mL volumetric flask by means of a glass funnel and dilute with distilled water to the 100 mL mark. Label this solution as the stock solution. Using a 1 mL graduated pipet, transfer a 0.5 mL sample of this stock solution into a 10 mL volumetric flask and dilute to the 10 mL mark with 0.02 M iron(III) chloride solution which has been buffered to a pH of 1.6 with hydrochloric acid-potassium chloride buffer. Place this solution in a test tube labeled solution A. In a similar fashion, prepare solutions labeled B, C, D, and E by diluting 0.40, 0.30, 0.20, and 0.10 mL aliquots of the sodium salicylate solution with the iron(III) chloride solution.

Observe the absorption of each violet solution spectrophotometrically at 530 nm utilizing the iron(III) chloride buffer solution as the reagent blank. The $FeCl_3$ solution should be used in a cuvette to set the spectrometer to 100% transmittance. Carry out the measurements on the Spectronic 20 or other similar instrument. Record all measurements in your laboratory notebook.

C. Determination of Aspirin in Commercial Tablets

On weighing paper, weigh about 0.16 g of a commercial aspirin tablet and transfer quantitatively to a 125 mL Erlenmeyer flask. Record the mass and brand of aspirin in your notebook. Prepare a stock solution in the same manner as was used for the pure acetylsalicylic acid. Only a 0.3 mL sample of the stock solution should be diluted with $FeCl_3$ solution (as was done to prepare solution C). This should be repeated with completely different aspirin tablets for successive trials. The presence of starch or fillers in the tablet may give the sodium salicylate solution a slightly milky or colored appearance, but this should disappear upon acidification. Using the Beer's law calibration curve, determine the percent of acetylsalicylic acid in the brand of commercial aspirin analyzed.

DISPOSAL

Unused NaOH and NaOH-aspirin solution: Put into a labeled waste container. Alternatively, you can add phenolphthalein and neutralize with 6 M H_2SO_4 or HCl (the first disappearance of pink color). Flush the resulting salt solution down the sink with running water.

Unused Fe-buffer and Fe-salicylate solutions: Put into a labeled waste bottle, which will be treated further for safe disposal.

Waste Solutions: All waste solutions should be put into a labeled waste container.

CALCULATIONS

Part B. Preparation of Standard Beer's Law Curve

1. Calculate the number of moles of acetylsalicylic acid used.
2. Calculate the molarity of the complex in the 100 mL stock solution.
3. Calculate the molarity of the complex in each solution prepared by dilution of the stock solution.
4. Plot the calculated concentration of the iron(III) salicylate complex versus the measured absorbance at 530 nm of each solution. Include this Beer's law plot in your report.
5. Calculate the molar absorptivity of the iron(III) salicylate complex, and be sure to include the correct units. This should be done by measuring the slope of the Beer's law plot.

Part C. Determination of Aspirin in Commercial Tablets

1. From the observed absorbance and the plot made in step 4 for known concentrations of aspirin, estimate the concentration of aspirin for each trial.
2. Calculate the mass of aspirin in the tablet.
3. Find the percentage of aspirin in the tablet.

Prelab   Name _____ Section _____

Spectrophotometric Analysis of Commercial Aspirin

1. State the mathematical equation known as Beer's law. Define each symbol.

2. What is to be done with the any unused reagent solutions?

3. What is the purpose of using an iron solution to dilute the aspirin sample?

4. If the absorbance of the aspirin-iron complex in Figure 12-4 is 1.50, what is the approximate molarity of the complex in the solution?

5. How is the absorptivity, *a*, determined from the Beer's law plot?

Report   Name _____ Section _____

DATA

B. 1. Preparation of Stock Solution

Mass of pure aspirin, g   _____

Moles of aspirin   _____

Concentration of aspirin in stock solution, M   _____

2. Dilution of Stock Solution and Absorbance Determinations

Solution	Concentration, M	%T	Absorbance
A	_____	_____	_____
B	_____	_____	_____
C	_____	_____	_____
D	_____	_____	_____
E	_____	_____	_____

Sample Calculations

Report   Name _____ Section _____

C. Determination of Aspirin in Tablets

Brand name _____

_____

	Sample 1	Sample 2	Sample 3
Mass of commercial sample, g	_____	_____	_____
%T of diluted solution	_____	_____	_____
Absorbance of diluted solution	_____	_____	_____
Concentration of aspirin in diluted solution, M	_____	_____	_____
Concentration of aspirin in commercial aspirin stock solution, M	_____	_____	_____
Moles of aspirin in commercial aspirin stock solution	_____	_____	_____
Mass of aspirin in sample, g	_____	_____	_____
Percent aspirin in tablet	_____	_____	_____
Average % aspirin		_____	

Sample Calculations

Report   Name _____ Section _____

QUESTIONS

1. What part of an absorption spectrum should be used for absorbance measurements–the peak of an absorption band, its side, or the valley between two bands?

2. The cells used in the absorbance measurement are calibrated accurately to ensure that the pathlength is as indicated. If a student measured the absorbance of a standard solution in a 1.00 cm cell and then *unknowingly* used a 1.01 cm cell for measurement of the unknown's absorbance, how would this affect the calculated concentration of aspirin in the unknown (increase, decrease, or no effect)? Explain.

3. Why is $FeCl_3$ solution used as the calibration blank rather than distilled water?

4. Has Beer's law been demonstrated by your experimental data? That is, is the absorbance versus concentration relationship linear? Explain.

13 Molecular Models and Covalent Bonding

INTRODUCTION

Much of the chemical behavior of matter can be related to the detailed structures and shapes of molecules. Structural theory is useful when we wish to relate experimental evidence to the more theoretical concepts of chemical bonding. The arrangement of atoms and ions in molecules and crystals is related to the distribution of bonding electrons within the structure.

Molecules are often represented by diagrams or molecular models in which balls or some other geometrical centers represent atoms and sticks or tubes represent the bonds between the atoms.

The purpose of this experiment is to use molecular models to help you understand more fully the theoretical concepts of covalent bonding and molecular structure.

Molecular models are designed to reproduce molecular structures in three-dimensional space. If models are correctly assembled, many subtle features concerning shapes of molecules (such as dipole moment, polarity, and bond angle) will become more clear.

One aspect of molecular structure is called *isomerism*. The most obvious type of structural isomerism is that in which there exists more than one way to assemble the atoms of a compound correctly. Sometimes only one of the possible structures is the stable form of a compound, but many times two or more structures are stable enough to exist in more than relatively small amounts. An example of a structural isomer is that of ethyl alcohol and dimethyl ether, shown in Figure 13-1. Note that both have the same molecular formula, C_2H_6O.

ethyl alchol   dimethyl ether

Figure 13-1. Two Structural Isomers of Formula C_2H_6O

Since many different kinds of models are used, it is impossible to give specific instructions here for each kit. One helpful rule of thumb in using ball and stick models is to select the appropriate ball for the central atom. For example, in methane, CH_4, the central carbon atom has four single bonds to the H atoms and no lone electron pairs. Thus the electron pair and molecular geometries are tetrahedral. It would be best to use a ball with four holes drilled tetrahedrally in it to represent the C atom. Each H atom requires a ball with only one hole, since each H has only one sigma bond and no lone pairs. Therefore, each H ball is connected to one hole in the C ball by a stick to represent the bond between C and H. The final assembly gives a three-dimensional model of the tetrahedral CH_4 molecule, in which all the bond angles are equal and the net effect of the C to H dipoles gives a nonpolar arrangement.

In general, therefore, for each atom choose a ball in which the number of holes equals the total number of sigma and pi bonds and lone electron pairs around that atom. Many elements are color-coded. If a given color is absent, use any ball that has the correct number and arrangement (such as trigonal planar, tetrahedral, or octahedral) of holes.

If multiple bonds are needed, you may need springs rather than sticks to connect the atoms. For lone pairs, simply insert a stick into the correct hole. Further instructions can be obtained from your laboratory instructor when necessary. Be sure to follow the instructions for the particular model kit that you are using.

General Rules for Predicting Molecular Geometries by Valence Shell Electron Pair Repulsion Theory (VSEPR)

1. Count the valence electrons in the entire molecule or ion.

2. Adjust this count for charge, if any, by adding (for anions) or subtracting (for cations) electrons.

3. Arrange atoms in the most symmetrical fashion. Frequently, the first element in the formula is the central atom in the structure.

4. Bond each terminal atom to the central atom.

5. Complete the octet on each terminal atom (except H).

6. Any unused electrons go on the central atom as unshared or lone pairs.

7. If the central atom is in periods 2 and 3, complete its octet, using multiple bonds if necessary.

8. Determine the molecular geometry based on mutual repulsion of electron pairs. Remember that nonbonding pairs are more repulsive than bonding pairs.

PROCEDURE

Students may work individually or in pairs.

Perform the following operations, when applicable, on each compound in the list provided by your instructor.

1. Draw the correct Lewis structure and any resonance forms.

2. Determine if any structural isomers exist.

3. Assemble a molecular model for the compound. Construct a model for each isomer from step 2. (Resonance structures require only one model.)

4. Sketch only one model, using a solid line for a bond in the plane of the paper, a wedge bond for a bond coming out of or in front of the plane of paper, and a dashed line for a bond in back of the plane of paper. For example, methane appears as

5. Indicate the following for each Lewis structure.

    a. Hybridization on central atom(s) (there may be more than one).

    b. Overall molecular geometry and electron pair geometry.

    c. Whether the molecule is polar or nonpolar. (Polar in this sense means that the species has a net dipole. Nonpolar means that there are no dipoles present, or that, if present, they cancel, resulting in no net dipole moment.)

14 Le Chatelier's Principle in Iron Thiocyanate Equilibrium

Due Next mon 1:25

INTRODUCTION

Chemical equilibrium represents a balance between forward and reverse reactions. Changes in experimental conditions–for example, concentration, pressure, volume, and temperature–disturb the balance and shift the equilibrium position so that more or less of the desired product is formed.

The general rule that is used to predict the direction in which an equilibrium will shift is known as *Le Chatelier's principle*. It states that if an external stress (i.e., change in concentration, pressure, volume, or temperature) is applied to a system at equilibrium, the system adjusts itself in such a way that the stress is partially offset.

$$N_2(g) + 3\,H_2(g) \rightleftharpoons 2\,NH_3(g) \qquad\qquad \Delta H = -\,92.6\ kJ \qquad\qquad [14.1]$$

In the above reaction at equilibrium, if N_2 and/or H_2 is added to the system, or if NH_3 is removed, the stress' will be relieved by forming more NH_3, and the equilibrium will shift to the right. If more NH_3 is added, the equilibrium will shift to the left producing more N_2 and H_2. If the pressure is increased (i.e., the volume decreased) the system will respond to reduce the pressure by shifting to the right. Note that there are 4 moles of gas on the left side of the equilibrium and only 2 moles of gas on the right side. If the mixture is heated, since it is an exothermic reaction, the system will shift to the left to consume the added energy.

In this experiment you will prepare an equilibrium system for the following reaction

$$Fe^{3+}(aq) + SCN^-(aq) \rightleftharpoons FeSCN^{2+}(aq) \qquad\qquad\qquad [14.2]$$

and study the response of this system to various stresses and explain these results in terms of LeChatelier's principle.

SAFETY PRECAUTIONS

Review the safety rules on pages 1 and 2.

All heavy metal ions and thiocyanate ions are toxic. Wash your hands with soap before you leave the lab.

PROCEDURE

Into a clean 400 mL beaker add 250 mL of deionized water, 1 mL of 1 M KSCN and 1 mL of 1 M $Fe(NO_3)_3$ solution. This stock solution will have an intense red color due to the formation of the complex $FeSCN^{2+}$. Obtain 10 clean medium-sized test tubes and label them 1 through 10. Add 10 mL of this stock solution into each one of the test tubes. Keep test tube 1 as a control and add the following reagents into test tubes 2 through 9 respectively:

20 drops = 1 mL

2   2 mL 1 M $Fe(NO_3)_3$
3   1 mL 1 M KSCN

4   0.5 mL (10 drops) 0.1 M $AgNO_3$
5   2 mL Conc. HCl
~~6   0.5 mL (10 drops) 0.1 M $Hg(NO_3)_2$~~
7   1 mL 1 M Na_3PO_4
8   1 mL 0.1 M $Na_2C_2O_4$
9   several crystals of solid NaF.

Cover each test tube with paraffin film and shake to mix. Heat test tube 10 in a hot water bath. Then compare the color intensity of each of the test tubes with that of the control test tube 1. The test tube with the darkest color has the highest concentration of $FeSCN^{2+}$. In each test record your observations and explain the results in terms of Le Chatelier's principle.

   Use the complexes and precipitates in Table 14-1 to explain your observations and write the equations of any reactions that occur.

Table 14-1. Complexes and Precipitates

	Ag^+	Hg^{2+}	Cl^-	PO_4^{3-}	F^-	$C_2O_4^{2-}$
Fe^{3+}:			$FeCl_4^-$	$FePO_4(s)$	FeF_6^{3-}	$Fe(C_2O_4)_3^{3-}$
SCN^-:	$AgSCN(s)$	$Hg(SCN)_4^{2-}$				

DISPOSAL

Dispose of any excess stock solution and the contents in test tubes 1 through 10 into a waste bottle.

Prelab   Name _____ Section _____

Le Chatelier's Principle in Iron Thiocyanate Equilibrium

    1. Write down the equilibrium constant expression for the following reaction:

$$Co^{2+}(aq) + 4\ Cl^-(aq) \rightleftharpoons CoCl_4^{2-}(aq)$$

    In which direction will the equilibrium shift if you (a) increase the concentration of Co^{2+} and (b) decrease the concentration of $CoCl_4^{2-}$?

    2. Using Le Chatelier's principle, predict the direction of the net reaction in each of the following equilibrium systems, as a result of increasing the pressure at constant temperature.

        a. $N_2(g) + O_2(g) \rightleftharpoons 2\ NO(g)$

        b. $PCl_5(g) \rightleftharpoons PCl_3(g) + Cl_2(g)$

        c. $CO(g) + Cl_2(g) \rightleftharpoons COCl_2(g)$

    3. What effect (shift to the right or left) does an increase in temperature have on each of the following systems at equilibrium?

        a. $3\ O_2(g) \rightleftharpoons 2O_3(g)$ $\Delta H = 284$ kJ

        b. $2SO_3(g) + O_2(g) \rightleftharpoons 2SO_3(g)$ $\Delta H = -198.2$ kJ

Do for Quz

Not part of Report

Report   Name _____ Section _____

Le Chatelier's Principle in Iron Thiocyanate Equilibrium

DATA

Test Tube Number	Reagent Added	Observation	Explanation
1		Orange-Red Clear	medium conc of FeSCN
2		Redder than #1	adding more reactant Shifts Right
3		redder than #1 Same as 2	adding more became darker shifts Righ
4		Pale Orange cloudy Percipitate at bottom	Ag & SNC react & form Sh a Percipitate
5		oranger clear	Shifts left forms FeCl₄ No Precipi
6		Slightly lighter than #1	Temp raise cause Shift left reactants endothermic
7		Yellow clear	Shifts left b/c removes ir from 1st reaction & forms Fe PO₄
8		Slight hue of yellow /clear	Shifts left No precip Complex Fe(C₂O₄)₃
9		nearly clear	

10

1st page

Quest 131 is 2nd page

15

A Kinetic Study of an Iodine Clock Reaction

INTRODUCTION

In general, for the chemical reaction

$$A + B \rightarrow products \qquad [15.1]$$

the *rate equation* may be expressed in the form

$$R = k[A]^m[B]^n = \frac{\Delta C}{\Delta t} \qquad [15.2]$$

In equation [15.2], R represents the rate of reaction [15.1] in terms of the increase in concentration of product(s), ΔC, divided by the time interval, Δt, required for the change–that is, $R = \Delta C/\Delta t$. Ordinarily, $\Delta C/\Delta t$ itself varies with time, so that rate measurements must specify either the initial rate (the rate over a very short time interval) or the average rate (over a longer time interval). k is a proportionality constant, known as the *specific rate constant*, a characteristic of each reaction at a specific temperature. The bracketed quantities [A] and [B] represent the molar concentrations of the reactants, A and B. The sum of the exponents m and n defines the **order** of the reaction. The order is determined experimentally, and cannot be deduced from the overall balanced equation for the reaction.

The order of a reaction is a key indicator of the *mechanism*–that is, the sequence of steps from reactants to products, or the pathway of the reaction. The slowest step in the mechanism is said to be *rate-controlling step* and has a *molecularity* (number of particles that collide) that must be equal to (or at least consistent with) the order of the overall reaction. For example, if the molecularity of the rate-controlling step is one, the overall reaction is first order; if the molecularity of the rate-controlling step is two, the overall reaction is second order, etc. Therefore, the order of the overall reaction provides this important insight into the stepwise mechanism of the reaction (the number of molecules that collide or in the case of a single molecule dissociate during the rate controlling step). It tells how the reaction proceeds and is an extremely useful quantity which aids our overall understanding of chemical kinetics.

In this experiment, the specific rate constant, k, and the order of reaction [15.3] will be determined by the *Method of Initial Rates*. In this method, the rates are observed for a series of reaction mixtures in which the concentration of one reactant is changed while that of the other(s) is held constant. The reaction that is being studied is that between the persulfate ion, $S_2O_8{}^{2-}$, and the iodide ion, I^-.

$$2\,I^- + S_2O_8{}^{2-} \longrightarrow I_2 + 2\,SO_4{}^{2-} \qquad [15.3]$$

The generalized rate equation for reaction [15.3] is given by the following equation.

$$R = k[I^-]^m[S_2O_8{}^{2-}]^n \qquad [15.4]$$

Reaction [15.3] proceeds at a rate that allows for convenient measurements of the rate equation. In order to help us detect the extent to which the reaction has taken place, an additional reaction (equation [15.5]) is carried out simultaneously in the same solution containing reaction [15.3].

$$I_2 + 2\,S_2O_3{}^{2-} \rightarrow 2I^- + S_4O_6{}^{2-} \qquad [15.5]$$

133

Compared with reaction [15.3], reaction [15.5] proceeds at a rate that is essentially instantaneous. The iodine, I_2, produced in reaction [15.3] is absorbed immediately by reaction with thiosulfate ion, $S_2O_3^{2-}$. As long as any $S_2O_3^{2-}$ remains in solution, the concentration of I_2 is effectively zero. When all of the $S_2O_3^{2-}$ is used up, however, the I_2 concentration will increase as reaction [15.3] continues. The presence of I_2 is detected by observing the deep blue color it forms with the starch indicator.

The combination of reactions [15.3] and [15.5] together with starch indicator constitutes one type of *Iodine Clock Reaction*. The "clock" or color change indicates when enough I_2 has been produced by reaction [15.3] to use up all of the $S_2O_3^{2-}$. A knowledge of the original $S_2O_3^{2-}$ concentration and the stoichiometric ratio between I_2 and $S_2O_8^{2-}$ leads to the quantity of $S_2O_8^{2-}$ that has reacted when the blue color appears. One mole of I_2 is produced in reaction [15.3] for each mole of $S_2O_8^{2-}$ reacted and reacts with 2 moles of $S_2O_3^{2-}$ during reaction [15.5]. Therefore, the quantity of $S_2O_8^{2-}$ that has reacted by the time the blue color appears is equal to one-half the quantity of $S_2O_3^{2-}$ present initially.

$$[S_2O_8^{2-}]_{reacted} = \frac{1}{2}[S_2O_3^{2-}]_{initial} \tag{15.6}$$

Since the initial concentration of $S_2O_3^{2-}$ is kept small in comparison with the quantities of I^- and $S_2O_8^{2-}$ used for reaction [15.3], the blue color will always appear before any appreciable quantity of $S_2O_8^{2-}$ is used up. Thus the concentrations of reactants and the rate in [15.4] will remain essentially constant during the time interval over which the rate is measured.

In Part A of this experiment, the study of reaction [15.3] will be conducted according to the conditions specified above and using the initial concentrations listed in Table 15-1. In each experiment the initial concentration of $S_2O_3^{2-}$ will be the same, while the concentration of either $S_2O_8^{2-}$ or I^- is doubled. In this way the order of reaction [15.3] with respect to each reactant can be determined. For example, if, while holding $[S_2O_8^{2-}]$ constant, the I^- concentration is doubled and this reduces by one-half the time needed for the blue color to appear (meaning that the rate doubles), then it may be concluded that the order of reaction [15.3] with respect to I^- is first order—i.e., $m = 1$. On the other hand, if doubling the concentration of I^- produces no effect on the rate, then it may be concluded that $m = 0$. Or, if doubling I^- results in the time needed for the appearance of the blue color to be one-fourth as long, the rate has quadrupled and the reaction is second order with respect to I^-—that is, $m = 2$. Similar conclusions may be drawn about the concentration-time dependency of $S_2O_8^{2-}$ and its order (n). This is called the *Method of Initial Rates*.

If, as a result of doubling the concentration of either I^- or $S_2O_8^{2-}$, the rate change is measurable but has not increased a whole number (integral) multiple (or close to it, certainly within experimental error), then it may be concluded that the reaction order with respect to that component is nonintegral. Under these circumstances, the order may be calculated according to the method used in the following example.

If, for example, the concentration of I^- is doubled while $[S_2O_8^{2-}]$ is held constant and if we define the initial concentrations of I^- as x and $2x$, the initial concentration of $S_2O_8^{2-}$ as y, the corresponding times are then t_1 (using x and y) and t_2 (using $2x$ and y).

Since the rates are inversely proportional to the times, $R_1 \propto 1/t_1$ and $R_2 \propto 1/t_2$, the corresponding rate expressions become

$$1/t_1 = k'(x)^m(y)^n \tag{15.7}$$

$$1/t_2 = k'(2x)^m(y)^n \tag{15.8}$$

Dividing equation [15.8] by equation [15.7] yields

$$\frac{t_1}{t_2} = \frac{(2x)^m}{(x)^m} \tag{15.9}$$

Equation [15.9] may be solved by taking its logarithm

$$\log \frac{t_1}{t_2} = \log \frac{(2x)^m}{(x)^m} = m \log 2$$

Solving for *m* we have

$$m = \frac{\log t_1/t_2}{\log 2} \tag{15.10}$$

An equation similar to [15.10] may be derived for *n*. These equations should be used whenever t_1 (or t_3) is not an integral multiple of t_2 (within ±5%).

In Part B, the effect of temperature on the rate of reaction [15.3] will be studied. With the same set of concentrations, the reaction will be carried out at several temperatures in addition to room temperature. From these results and equation [15.11] the *activation energy*, E_A, for the reaction may be determined.

$$\log_{10} k = -\frac{E_A}{2.3RT} + \text{constant} \tag{15.11}$$

In equation [15.11] *k* is the rate constant at each of the different temperatures, E_A is the activation energy in J mol^{-1}, *R* is the gas law constant, 8.314 J mol^{-1} K^{-1}, and *T* is the absolute temperature, K.

Since the reactions at the various temperatures (Table 15-2) all involve the same set of concentrations, the rate constants (k_1 and k_2) at two different temperatures (T_1 and T_2) will have the same ratio as the observed rates, which are inversely proportional to the times. Thus, if the log of the relative rates (100/time) is plotted as a function of $1/T$, the activation energy, E_A, can be obtained from the slope ($-E_A/2.303R$) of the resulting straight line.

SAFETY PRECAUTIONS

Review the safety rules on pages 1 and 2.

Be careful in handling these chemicals. If spills occur, rinse with water and wipe thoroughly.

When pipeting, use a suction device. **Never pipet by mouth!**

Be careful in handling the thermometer to avoid breakage. Hg is toxic, and broken glass can cause cuts.

PROCEDURE

A. Dependence of Reaction Rate on Concentration

For this investigation, gathering of data is accomplished more conveniently if students work in pairs. The three experiments are to be carried out at room temperature, which should be recorded in the laboratory notebook and included in the report for this experiment.

The volumes of the KI, $(NH_4)_2S_2O_8$, and $Na_2S_2O_3$ solutions should be measured with a pipet. The volumes of the KCl and $(NH_4)_2SO_4$ solutions do not need to be known as precisely and can be measured with a graduated cylinder. The latter solutions are used rather than water in diluting to a constant volume so that the total ionic strength of the reaction mixture, which has some effect on the reaction rate, can be kept approximately constant. Use clean, dry, labeled medium-sized test tubes to obtain 7 to 8 mL of each solution to take back to your lab bench.

Table 15-1. Quantities of reactants used at room temperature

Reactant	Volume of Reactant, mL		
	Expt. 1	Expt. 2	Expt. 3
0.200 M KI	1.00	2.00	2.00
0.200 M KCl	1.00	0	0
0.100 M $(NH_4)_2S_2O_8$	2.00	2.00	1.00
0.100 M $(NH_4)_2SO_4$	0	0	1.00
0.005 M $Na_2S_2O_3$	1.00	1.00	1.00

Handwritten annotations:
KI, KCl, Na₂S₂O₃ — grad cyl — Tube 1
(NH₄)₂S₂O₈, (NH₄)₂SO₄ — grad cyl — Tube 2

In each experiment, pipet the specified volume(s) of KI (and KCl) solution(s) into a small test tube, which will be used as the reaction container. Pipet 1.00 mL of 0.005 M $Na_2S_2O_3$ into this tube and add 2 drops of starch solution. Insert a thermometer into the reaction container. Pipet the specified volume(s) of $(NH_4)_2S_2O_8$ [and $(NH_4)_2SO_4$] into a separate test tube.

Observe the time (sweep second hand on watch or wall clock) while pouring the persulfate solution from the test tube into the reaction tube. Swirl the solution so as to mix thoroughly. Record the time at which the solutions were mixed and the time required to turn the solution blue. Constant observation is necessary because the blue color will appear suddenly. Record the temperature of the solution.

Rinse the tubes thoroughly between experiments and repeat each one to reproduce the elapsed time to within 5 sec.

B. Dependence of Reaction Rate on Temperature

Carry out reaction [15.3] at the temperatures specified in Table 15-2. Use the same concentrations as in Experiment 2 of Table 15-1.

Table 15-2. Iodine Clock Reaction and Temperature

Experiment	Temperature, °C
2	Room temperature (r. t.)
4	10° above r. t.
5	10° below r. t.
6	About 0°C or 20° below r. t.

Instead of mixing at room temperature, place the two test tubes in a beaker of water heated with a burner or cooled with an ice-water bath to the desired temperature. Place a thermometer in the reaction tube. After several minutes at the specified temperature, mix the two solutions by pouring the solution from the persulfate test tube into the reaction tube, which is kept in the water bath. Swirl the tube as before. Record the times of mixing and when the color change occurs and the temperature at the time of the color change. Repeat these experiments if time permits.

DISPOSAL

All the solutions of reaction products are classified as nonhazardous and may be flushed down the sink with running water. Unused reactant may be disposed in a labeled waste container.

Report

Name _____ Section _____

Lab Partner _____

DATA

A. Dependence of Reaction Rate on Concentration

Temperature = _23___ °C = _____ K

Monica Briscoe
Brenden Whitaker

Account for dilution

Experiment	Initial Concentrations, M [I⁻]	[$S_2O_8^{2-}$]	Elapsed Time, sec	Relative rate, 100/Time		
1	0.040	0.040	71	1.19	average	84
			97 sec			
2			38 sec	0.901	average	111
			73 s			
3			75 sec	0.954	average	180.5
			105.5			

Give the exact number below, unless it is within 5% of a whole number.

Order of reaction with respect to I⁻ _____ m

Order of reaction with respect to $S_2O_8^{2-}$ _____ n

Overall order of reaction _____

B. Dependence of Reaction Rate on Temperature

Experiment	Temperature, T, K	1/T	Elapsed Time, sec	Relative Rate 100/Time	Log of Relative Rate
2	296	0.00338	38 sec	2.63	0.419
4	206 c°	0.00327	15 sec	6.67	0.824
5	285.2°	0.00351	160	0.625	-0.204
6	276.15	0.00362	215 s	0.465	-0.333

★Plot (log of relative rate) versus $\frac{1}{T}$ on graph paper (or use a spreadsheet to get a linear least squares plot) and include it with your report. Calc Ea!

Sample Calculations

Report

Name _____ Section _____

Lab Partner _____

QUESTIONS

1.   From your results in Part A, can a prediction be made about the nature of the rate-controlling step in this reaction? If not, why not? If so, give a plausible chemical equation for this step. (*Hint:* Think about the connection between the molecularity of the rate-controlling step and the order of the overall reaction.)

2.   From your results in Part A, calculate the time for the blue color to appear for this system if 1.5 mL KI, 1.5 mL KCl, 1.0 mL $(NH_4)_2S_2O_8$, and 1.0 mL $Na_2S_2O_3$ from Table 15-1 are mixed.

3.   Using your linear plot of **log(relative rate)** versus $\frac{1}{T}$ in Part B, measure the experimental slope and use it to calculate the energy of activation, E_A, in units of kJ mol^{-1}.

4.   Use your data from experiments 2, 4, and 5 and the Arrhenius equation

$$\log \frac{k_2}{k_1} = \frac{E_A}{2.303\ R}\left[\frac{T_2 - T_1}{T_2 T_1}\right]$$

to calculate the value of E_A for this reaction. How does this value compare with that obtained from the slope of the plot in question 3? Which one do you think is more accurate? Explain.

5.   Do your data agree with the postulate that the reaction rate will double for a 10°C rise in temperature? Support your answer with a calculation.

Prelab

Name _____ Section _____

Lab Partner _____

A Kinetic Study of an Iodine Clock Reaction

1. For all experiments in Part A, how many millimoles of persulfate ion have reacted by the time the blue color appears?

2. In addition to the effects of catalysts, what other factors can influence the rate of a chemical reaction?

3. What is meant by the term "rate-controlling step" of a chemical reaction?

4. How is the molecularity of the rate-controlling step related to the overall reaction order?

5. How will a decrease in temperature affect the rate of a reaction (increase, decrease, no effect)? Explain.

6. From the data below calculate the rate law expression for the reaction of A with B.

Experiment	[A]	[B]	time, s
1	0.100 M	0.100 M	30
2	0.200 M	0.100 M	15
3	0.200 M	0.050 M	60

16 Determination of an Equilibrium Constant

INTRODUCTION

A system is considered to be in a state of equilibrium when its properties do not change as time passes. For chemical systems, this means that all chemical forces are in balance and that all the physical properties of the system, such as color, density, and the concentrations of all chemical species remain constant. An important way of measuring the extent to which a reaction has proceeded once equilibrium is established is to measure the equilibrium constant for the reaction. For the generalized reaction

$$a\,A + b\,B \rightleftharpoons c\,C + d\,D \qquad\qquad [16.1]$$

the equilibrium constant expression is defined as

$$K = \frac{[C]^c[D]^d}{[A]^a[B]^b} \qquad\qquad [16.2]$$

where [A], [B], [C], and [D] are the molar concentrations of the respective components *at equilibrium*. The concentration terms in [16.2] are raised to powers (exponents) equal to their stoichiometric coefficients. K is constant under all conditions except for changes in temperature. In this experiment, we shall evaluate K for a specific reaction under different initial concentration conditions to test its constancy at constant temperature.

In order to determine the equilibrium constant, the stoichiometry of the chemical reaction must be known. In addition, we must have an analytical method for measuring the concentrations of reactants and products at equilibrium. Alternatively, it is possible to determine the value of the equilibrium constant by measuring the equilibrium concentration of only one of the species involved in the reaction, provided that the initial concentrations of all species are known.   In this experiment the second method will be used.   The concentration at equilibrium of one of the reaction products will be measured by titration, and this value will be used to calculate the equilibrium concentrations of the remaining species. Once the equilibrium concentrations of all the reaction species are known, the equilibrium constant will be calculated.

In this experiment we shall study the following equilibrium system.

$$\underset{\substack{\text{ethyl acetate}\\\text{(EtAc)}}}{CH_3CH_2O\text{--}\overset{\displaystyle O}{\overset{\displaystyle \|}{C}}\text{--}CH_3} \;+\; \underset{\text{water}}{H_2O} \;\rightleftharpoons\; \underset{\substack{\text{ethyl alcohol}\\\text{(EtOH)}}}{CH_3CH_2\text{--}OH} \;+\; \underset{\substack{\text{acetic acid}\\\text{(HAc)}}}{CH_3COOH}$$

$$[16.3]$$

The reaction may be represented more simply by using abbreviated names.

$$EtAc + H_2O \rightleftharpoons EtOH + HAc \qquad\qquad [16.4]$$

From equation [16.4], the equilibrium constant K_c is defined by the following expression.

$$K_c = \frac{[EtOH][HAc]}{[EtAc][H_2O]} \qquad\qquad [16.5]$$

If the initial concentrations of all reaction species are known, the determination of the equilibrium concentration of acetic acid will permit us to calculate the equilibrium constant for this reaction.

Note: Normally the concentration of a pure liquid such as water does not appear in the equilibrium expression because its concentration does not vary significantly during the course of the reaction. However, in this experiment water is present in a concentration that does change as the reaction progresses and therefore must be included in the equilibrium expression.

SAFETY PRECAUTIONS

Review the safety rules on pages 1 and 2.

Be careful in handling these chemicals. If spills occur, rinse with water and wipe thoroughly.

If you spill a chemical on you, wash with soap and lots of running water. Inform your instructor.

PROCEDURE

Part A

1. **Solution preparation:** Obtain three clean and dry 125- or 250-mL Erlenmeyer flasks. Mark them A, B, and C. Add the following materials to each flask using a buret or pipet. These three solutions will be used to make three independent determinations of the equilibrium constant.

   A. 5.00 mL 3 M HCl + 5.00 mL of ethyl acetate

   B. 5.00 mL 3 M HCl + 4.00 mL of ethyl acetate + 1.00 mL of distilled water 5 mL

   C. 5.00 mL 3 M HCl + 3.00 mL of ethyl acetate + 2.00 mL of distilled water

   Notice that the volumes must be measured as precisely as possible. If you do not add exactly the specified volume of a reagent, write down in your laboratory notebook the exact volume that you do add and use this volume in your calculations. The HCl must be present even though it does not appear in the balanced chemical equation because this reaction is acid-catalyzed; the rate at which the reaction reaches equilibrium is greatly increased by the presence of acid. Stopper each flask tightly and swirl the contents vigorously. Allow the flasks to stand at room temperature until your next laboratory period. It will take about one day for equilibrium to be achieved.

   Start here

2. **Determination of moles of HCl and H_2O in 3 M HCl solution:** Review proper titration technique described on page 66 in Experiment 7. Using a graduated cylinder, add approximately 20 mL of distilled water and three drops of phenolphthalein to 5.00 mL of the same 3 M HCl solution used as the catalyst in the preparation of the equilibrium solutions. Titrate the solution with 1.00 M sodium hydroxide solution. Use care in titrating; the acid and base are more concentrated than is typical of many titrations and this can cause the end point to arrive suddenly.

3. Repeat the above titration until two successive titrations have a relative average deviation of 1% or less.

Part B

1. **Determination of total moles of acid at equilibrium:** After equilibrium has been established in the reaction flasks, add approximately 20 mL of distilled water to increase the volume and titrate each solution (A, B, and C) with 1.00 M NaOH; use phenolphthalein as the indicator. This 20 mL volume of water is *not* included in your calculations. There is not time enough for it to react.

   read on bottle

2. Have your instructor demonstrate the technique of detecting the odor of a reaction. Only after you have seen that demonstration, and then only with the explicit permission of your instructor, very cautiously smell the reaction mixture just after you have prepared it and then, at least a day later, just before you titrate it. Can you tell that a reaction has occurred?

CALCULATIONS

Normally the concentrations of all substances at equilibrium must be used to calculate the equilibrium constant. However, in this experiment the number of moles of each substance at equilibrium may be used instead of concentration. This is because in the equilibrium expression shown in [16.5] all volume terms in molarity will cancel algebraically.

In order to determine the equilibrium constant, the *initial* number of moles of EtAc, H_2O, HAc, and EtOH must be calculated. Since only EtAc and water are present initially, the initial number of moles of the products is zero. The initial number of moles of EtAc and water will be calculated using their measured volumes, molar masses, and densities. Since the HCl catalyst was added as an aqueous solution, the amount of water present in this solution must be included in the calculation of the initial moles of water. (The density of 3 M HCl = 1.05 g/ml.) The number of moles of HAc at equilibrium is calculated using the titration data for the moles of HCl present and total moles of acid present in the equilibrium solution. Once the initial moles of all species and the equilibrium moles of HAc have been determined, the equilibrium moles of EtAc, EtOH, and water may be calculated using the stoichiometry of the chemical reaction. The equilibrium constant is calculated using the equilibrium moles of HAc, EtOH, EtAc, and water in equation [16.5]. An example of the type of calculation to be performed is shown below.

EXAMPLE: A student prepared two flasks. Flask A contained 5.00 mL of 6 M HCl (density = 1.11 g/mL). Flask B contained 5.00 mL of 6 M HCl, 2.00 mL of ethyl acetate (density = 0.893 g/mL), and 3.00 mL of water. Flask A required 28.90 mL of 1.00 M NaOH for titration; flask B required 39.12 mL of 1.00 M NaOH. Calculate K_c. *Note*: in *your* experiment you are using 3M HCl, not 6 M.

1. Use your titration data from flask A to calculate the moles of HCl in 5.00 mL of 6 M HCl solution. This is necessary since the molarity of HCl is known to only one significant figure. Use the volume and molarity of NaOH to find the moles of NaOH used to titrate the HCL in flask A. This is equal to the number of moles of HCl present in both flask A and flask B.

    $$0.02890 \text{ L} \times 1.00 \text{ M} = 0.0289 \text{ mol NaOH} = 0.0289 \text{ mol HCl}$$

    12.9   0.0129 L × 0.991

2. Calculate the moles of water in 5.00 mL of HCl solution. The HCl solution is made up of the solute, HCl, and the solvent, H_2O. The total mass of the HCl solution is calculated using the volume and density of the solution.

    $$5.00 \text{ mL} \times 1.11 \text{ g/mL} = 5.55 \text{ g of HCl solution}$$

    The mass of *pure* HCl in the HCl solution is calculated using the moles of HCl dissolved in the solution and its molar mass.

    $$0.0289 \text{ mol} \times 36.5 \text{ g/mol} = 1.06 \text{ g HCl in the HCl solution}$$

    The difference between the total mass of solution and the mass of solute (HCl) is the mass of water.

    $$5.55 - 1.06 = 4.49 \text{ g of water}$$

    The number of moles of water is calculated using the molar mass of water.

    $$4.49 \text{ g} / (18.0 \text{ g/mol}) = 0.249 \text{ mol of } H_2O$$

3. Calculate the initial moles of water in flask B. The number of moles of water in flask B is equal to the sum of moles of water in the 5.00 mL of HCl solution and in the 3.00 mL of water added to flask B.

    $$3.00 \text{ mL} \times 1.00 \text{ g/mL} = 3.00 \text{ g of water in 3 mL of water added}$$
    $$3.00 \text{ g}/(18.0 \text{ g/mol}) = 0.167 \text{ mol of water in 3 mL of water added}$$
    $$0.167 \text{ mol} + 0.249 \text{ mol} = 0.416 \text{ total mol of water in flask B}$$

4. Calculate the initial moles of EtAc in flask B. Use the density and volume to find the mass of EtAc and then calculate the moles using molar mass (density EtAc = 0.893 g/mL).

    $$2.00 \text{ mL} \times 0.893 \text{ g/mL} = 1.79 \text{ g of EtAc}$$
    $$1.79 \text{ g} / (88.0 \text{ g/mol}) = 0.0203 \text{ initial mol of EtAc}$$

5. Calculate the moles of HAc at equilibrium. First calculate the total moles of acid in flask B from the titration data. Then subtract the moles of acid due to the HCl solution.

    $$0.03912\text{L NaOH} \times 1.00\text{M NaOH} = 0.0391 \text{ mol NaOH} = 0.0391 \text{ mol acid}$$
    $$0.0391 \text{ mol acid} - 0.0289 \text{ mol HCl} = 0.0102 \text{ mol HAc}$$

0.991 M NaOH

6. Calculate the moles of EtAc, EtOH, and H_2O at equilibrium. For reactants, the moles used during the reaction are subtracted from the initial moles. For products, the moles formed in the reaction are added to the initial moles. The moles reacted or produced during the reaction are based on the moles of HAc formed and the stoichiometry of the reaction. In this reaction the stoichiometry is a one-to-one relationship for all the species. These calculations are summarized below:

	EtAc(aq)	+	H_2O(soln)	⇆	EtOH(aq)	+	HAc(aq)
Initially:	0.0203 mol		0.416 mol		0.00 mol		0.00 mol
Changes:	−0.0102 mol		−0.0102 mol		+0.0102 mol		+0.0102 mol
At equilibrium:	0.0101 mol		0.406 mol		0.0102 mol		0.0102 mol

The value of the equilibrium constant is calculated as follows.

$$K_c = \frac{[HAc][EtOH]}{[EtAc][H_2O]} = \frac{(0.0102)(0.0102)}{(0.0101)(0.406)} = 0.0254$$

Note that the student in the example used 6 M acid rather than 3 M acid and also performed the experiment at a different temperature so that his or her value of K_c will differ from yours. This example is not meant to be followed exactly during the report write-up, but it is intended merely to demonstrate the general method of performing the calculations.

DISPOSAL

Unused HCl and NaOH: There should be no reagents left over if burets are used to dispense them. However, if unused acid and base remain, mix them together in a beaker, add phenolphthalein, and neutralize with 1 M HCl (the first disappearance of pink color) or 1 M NaOH (first appearance of permanent pink color). Flush the resulting salt solution down the sink with running water.

Titration solutions: These solutions already are neutral, and any unreacted ethyl acetate will hydrolyze to nonhazardous compounds. Therefore, they may be safely flushed down the sink with running water.

Hooman Saberinia
hoomaneee

Prelab   Name _____ Section _____

Determination of an Equilibrium Constant

1. Write an equilibrium constant expression for the chemical reaction of ammonia with water. (In this reaction assume that the concentration of water will remain constant.)

$$NH_3(aq) + H_2O(l) \leftrightarrows NH_4^+(aq) + OH^-(aq)$$

$$K = \frac{[NH_4][OH]}{[NH_3][H_2O]}$$

2. a. Why is HCl added to the reaction mixture?

   The reaction is acid-catalyzed ; HCl will increace the rate the reaction reaches equilibrium

   b. Does the number of moles of HCl change from the initial to equilibrium conditions? If yes, how? If no, why not?

   Do not change becauce HCl is a catalyst

3. In our experiment, why must the concentration of water be included in the equilibrium expression?

look over
good Quiz
Quest.

4. A flask containing 5.00 mL of 3 M HCl solution requires 14.45 mL of 1.00 M NaOH for titration. How many moles of HCl are present in the solution? The density of 3 M HCl is 1.05 g/mL.

   0.01445 × 1 M = 0.01445 mol

5. If you have some unused HCl when you are finished, how must it be disposed?

   Add NaOH or another base, Flush salt down sink with running water

17 Determination of Iron in Vitamins

INTRODUCTION

Vitamin supplements contain a wide variety of vitamin and mineral nutrients. Almost all tablets contain some iron, which is an element of critical importance to the human circulatory system. For example, the uptake of oxygen and the removal of carbon dioxide from our blood depend on the hemoglobin molecule, at the center of which is an atom of iron. A deficiency of iron in the blood can lead to a loss of hemoglobin and the onset of certain types of anemia. Severe cases may require the intake of iron supplements in the diet.

In this experiment a vitamin tablet containing iron is digested in strong acid, and the cooled mixture is filtered into a volumetric flask. After making the appropriate dilutions and adjusting the pH, the dissolved iron is reduced to iron(II) with hydroquinone, as shown by equation [17.1]. Next, it is complexed with *o*-phenanthroline to form a highly colored, red-orange complex ion (equation [17.2]). The concentration of the complex is determined spectrophotometrically by comparison of its absorbance to that of a series of standard solutions. The total amount of iron in the tablet is calculated from the concentration of the complex.

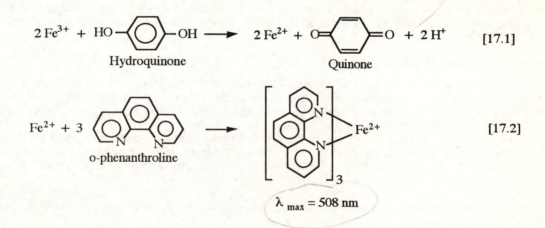

$$[17.1]$$

$$[17.2]$$

$$\lambda_{max} = 508 \text{ nm}$$

Important! A discussion of the theory and practice of spectrophotometry and Beer's law is given in Experiment 12 of this manual. Before beginning this experiment, be sure to read these pages.

SAFETY PRECAUTIONS

Review the safety rules on pages 1 and 2.

Work in the fume hood when digesting the vitamin tablet.

Wear eye protection at all times. HCl solutions are harmful to eyes and skin, especially when hot. If any gets on your skin, wash thoroughly with running water.

PROCEDURE*

A. Sample tablet extraction , filtration, and dilution

Obtain a vitamin tablet containing iron and record the label value of the mg Fe per tablet. Pour 25 mL of 6 M HCl into a clean 100- or 150-mL beaker. Add the vitamin tablet, cover with a watch glass, and **(in a fume hood)** heat gently for 15 minutes. Cool and filter or centrifuge the solution directly into a 100-mL volumetric flask, washing the beaker and filter several times with distilled water to complete the quantitative transfer. Dilute to the mark with deionized water, and mix well. Label this original sample as *solution 1*. Dilute the original solution by pipeting 5.00 mL (use 10.00 mL if the label value for iron is less than 15 mg) to 100 mL in a clean volumetric flask. Label this flask *solution 2*.

B. Standard Iron Solutions

Pipet 5.00 mL of stock Fe standard (40.0 ppm) into a clean 50 mL beaker or a test tube. Add several drops of methyl orange indicator solution. At this point the indicator should have a red color. Add sodium citrate reagent drop-wise, with mixing, until the red color changes completely to orange. Record the number of drops used. (If you are not sure of the color change, try this step with 5 mL of pure water.)

Pipet 5.00 mL of stock Fe standard (40.0 ppm) into a clean 50 mL volumetric flask and add the same number of drops of sodium citrate reagent solution as was used in the previous step. Add 1 mL of hydroquinone reagent, 1.5 mL of o-phenanthroline solution, dilute to the mark with water, and mix well. Label this as standard solution A.

Prepare three more working standards by pipeting 3.00, 1.00, and 0.500 mL of stock Fe standard, respectively, into three 50 mL volumetric flasks. However, add sodium citrate reagent in proportion to the volume of Fe solution. (For example, if 5 mL of standard required 15 drops of reagent, 3 mL will need 9 drops.) . To each add 1 mL of hydroquinone reagent, 1.5 mL of o-phenanthroline solution, dilute to the mark with water, and mix well. Label these three standard solutions as B, C, and D, respectively.

Prepare a blank solution in a clean 50 mL volumetric flask by adding 5 drops of citrate reagent, 1 mL of hydroquinone reagent, and 1.5 mL of o-phenanthroline solution. Dilute to the mark with distilled water and mix well.

C. Sample Preparation for Measurement

Pipet 5.00 mL of the diluted vitamin sample (*solution 2*) into a clean 50 mL beaker or a test tube. Add several drops of methyl orange indicator solution. At this point the indicator should have a red color. Add sodium citrate reagent dropwise, with stirring, until the red color changes completely to orange. Record the number of drops used. (This could take a lot more than the standard solution required.)

Into a 50-mL volumetric flask, pipet 5.00 mL of the **diluted** vitamin sample (*solution 2*). Add the required volume of citrate reagent, as used in the previous step. Then add 1.0 mL of hydroquinone, and 1.5 mL of o-phenanthroline solution. Dilute to the mark with water and mix well. Label it *solution 3*. (Note that this step dilutes the sample further by 10 times.)

Color development is moderately slow, so allow all solutions to stand for *at least 10 minutes* before measuring their absorbances.

D. Absorbance Measurements

Set your instrument's wavelength to 508 nm. As your instrument requires, set the zero %T with no light going through the cell compartment and use the blank solution to set the %T = 100 (or A = 0). Measure the percent transmittance or absorbance of each standard solution and of *solution 3* . Be sure to set the zero and 100 %T setting for each measurement. If %T is used, calculate the Absorbance (A) of each solution from the equation

$$A = 2.000 - \log(\%T) \qquad\qquad [17.3]$$

DISPOSAL

All solutions may be disposed in the same waste bottle labeled *inorganic waste solutions*.

* **Reference**: Adapted from R.C. Atkins, *J. Chem. Ed.,* **52**, 550 (1975).

Due Week from tomorrow
lot 5 tues
can write Jointly

Quiz
Statement at top saying Joint effort
w/signiture

18 Weak Acids, Weak Bases, and their Salts

INTRODUCTION

The purpose of this experiment is to determine the pH of various concentrations of acids, bases, and salts. The pH values are determined by one of three methods:

1. Matched indicators

2. pH paper

3. pH meter

These data will be used to determine the acid–base dissociation constants of the substances being investigated. The variation of the degree of ionization of a weak acid and of a weak base with concentration also will be determined.

A. Strong Acids and Bases

When a strong acid such as HCl or a strong base such as NaOH is dissolved in water, it dissociates completely into ions as shown in the following two equations.

$$HCl(g) + H_2O \rightleftharpoons H_3O^+(aq) + Cl^-(aq) \qquad [18.1]$$

$$NaOH(s) \xrightarrow{\text{H}_2\text{O}} Na^+(aq) + OH^-(aq) \qquad [18.2]$$

As a result, the concentration of hydronium ions, $[H_3O^+]$, or the concentration of hydroxide ions, $[OH^-]$, essentially is equal to the concentration of the strong acid or the strong base, respectively.

Hydronium ion concentrations commonly range from 1.0 M to 10^{-14} M and frequently are expressed on the logarithmic scale of pH.

$$pH = -\log[H_3O^+] \qquad [18.3]$$

Furthermore, $[OH^-]$ is related to $[H_3O^+]$ by the ion product of water, K_w, which at 25°C has the value

$$K_w = [H_3O^+][OH^-] = 1.0 \times 10^{-14}. \qquad [18.4]$$

The relationship between pH and pOH $(-\log[OH^-])$ is obtained from equation [18.4].

$$-\log K_w = -\log[H_3O^+] - \log[OH^-] = +14.00$$

$$pK_w = pH + pOH = 14.00 \qquad [18.5]$$

Consequently, if one of pH, pOH, $[H_3O^+]$, or $[OH^-]$ is known, the other three may be calculated by using equations [18.3], [18.4], and [18.5].

157

B. Weak Acids and Bases

In contrast to strong acids and bases, weak acids and bases do not dissociate completely in solution. They ionize only partly, as shown for nitrous acid (HNO_2) and hydrazine (N_2H_4) in the following equations.

$$HNO_2(aq) + H_2O \leftrightarrows H_3O^+ (aq) + NO_2^-(aq) \qquad [18.6]$$

$$N_2H_4(aq) + H_2O \leftrightarrows OH^-(aq) + N_2H_5^+(aq) \qquad [18.7]$$

The equilibrium constants for these reactions are known as dissociation or ionization constants. For a given weak acid or weak base, the ionization constant has a definite value at a specific temperature. The ionization constant expressions and their values at 25°C for nitrous acid and hydrazine are

$$K_i = K_a \text{ (HNO}_2) = \frac{[H_3O^+][NO_2]}{[HNO_2]} = 4.5 \times 10^{-4} \qquad [18.8]$$

$$K_i = K_b \text{ (N}_2\text{H}_4) = \frac{[OH^-][N_2H_5^+]}{[N_2H_4]} = 9.8 \times 10^{-7} \qquad [18.9]$$

K_a is an acid ionization constant, and K_b is a base ionization constant.

Ionization constants are determined experimentally by a method very similar to that used in this experiment. The pH of a solution of weak acid or base of known concentration is determined, and these data are used to calculate the value of the ionization constant.

EXAMPLE: The pH of a 1.0 M solution of benzoic acid, $HC_7H_5O_2$, is 2.11.   Calculate the ionization constant of this benzoic acid.

A pH of 2.11 is equivalent to $[H_3O^+] = 0.0078$ M.   The ionization reaction follows, and the initial and equilibrium concentrations are given below each species.

	$HC_7H_5O_2$	+ H_2O	$\leftrightarrows$	$C_7H_5O_2^-$	+ H_3O^+
Initial:	1.00 M				
Equilibrium:	(1.00 - 0.0078) M			0.0078 M	0.0078 M

Since only equilibrium concentrations are used in the ionization constant, K_a becomes

$$K_a = \frac{[H_3O^+][C_7H_5O_2^-]}{[HC_7H_5O_2]} = \frac{(0.0078)(0.0078)}{(1.00 - 0.0078)} = 6.1 \times 10^{-5} \qquad [18.10]$$

leave out

Because weak acids and bases only partially ionize, the degree of ionization, α, of a weak acid or base often is calculated. For example, consider a weak base B, which ionizes as shown below.

$$B(aq) + H_2O \leftrightarrows BH^+(aq) + OH^-(aq) \qquad [18.11]$$

$$\alpha = \frac{[BH^+]_{eq}}{[B]_{init}} = \frac{[OH^-]_{eq}}{[B]_{init}} \qquad [18.12]$$

The subscript "eq" indicates equilibrium concentration and the subscript "init" indicates initial concentration. While the ionization constant of a given acid or base depends only on temperature, the degree of ionization also varies inversely with concentration.

C. Salts of Weak Acids and Bases

Although soluble salts dissociate completely into their ions when they dissolve in water, the fate of these ions depends on the acid or base from which they are derived. The anions of strong acids and the cations of strong bases do not react with water except to become hydrated—surrounded by water molecules. On the other hand, the anions and cations of weak acids and bases react with water by a process sometimes known as *hydrolysis*.

Thus the salt sodium nitrite, $NaNO_2$, first dissociates completely in water

$$NaNO_2 \xrightarrow{H_2O} Na^+(aq) + NO_2^-(aq) \qquad [18.13]$$

and then the nitrite ion acts as a weak base, accepting a proton from water to form its conjugate acid, HNO_2.

$$NO_2^-(aq) + H_2O \rightleftharpoons HNO_2(aq) + OH^-(aq) \qquad [18.14]$$

The equilibrium constant, K_b, for equation [18.14] is

$$K_b = \frac{[HNO_2][OH^-]}{[NO_2]} \qquad [18.15]$$

Values of K_b for anions are rarely tabulated since they may be calculated from the ionization constants of the corresponding conjugate acids. Algebraic manipulation of equation [18.15] shows the relationship between K_a and K_b of a conjugate acid–base pair.

$$K_b = \frac{[HNO_2][OH^-]}{[NO_2]} \times \frac{[H_3O^+]}{[H_3O^+]}$$

$$K_b = \frac{[HNO_2]}{[H_3O^+][NO_2]} \times [H_3O^+][OH^-]$$

$$K_b = \frac{1}{K_a} \times K_w = \frac{K_w}{K_a} \qquad \text{or} \qquad K_w = K_a \times K_b \qquad [18.16]$$

For the salt hydrazine hydroiodide, N_2H_5I, the equations below are analogous to [18.13]-[18.16].

$$N_2H_5I \xrightarrow{H_2O} N_2H_5^+(aq) + I^-(aq) \qquad [18.17]$$

$$N_2H_5^+(aq) + H_2O \rightleftharpoons N_2H_4(aq) + H_3O^+(aq) \qquad [18.18]$$

$$K_a = \frac{[N_2H_4][H_3O^+]}{[N_2H_5^+]} \qquad [18.19]$$

$$K_a = \frac{K_w}{K_b} = [H_3O^+][OH^-] \times \frac{[N_2H_4]}{[N_2H_5^+][OH^-]} \qquad [18.20]$$

The pH and concentration of a salt dissolved in water are the data necessary to determine the values of K_a and K_b for a weak conjugate acid–base pair.

EXAMPLE: A 1.00 M solution of sodium benzoate, $NaC_7H_5O_2$, has a pH of 9.11. What is the value of K_b for the benzoate anion (the conjugate base of benzoic acid) and the acid ionization constant of benzoic acid?

A pH = 9.11 is equivalent to pOH = 4.89 or $[OH^-] = 1.3 \times 10^{-5}$ M. The hydrolysis reaction follows.

	$C_7H_5O_2^-$	+	H_2O	$\leftrightarrows$	$HC_7H_5O_2$	+	OH^-
Initial:	1.00 M						
Equilibrium:	$(1.00 - 1.3 \times 10^{-5})$ M ≈ 1.00 M				1.3×10^{-5} M		1.3×10^{-5} M

Then K_b for benzoate anion and K_a or benzoic acid can be determined. Note that for any conjugate acid–base pair, $K_aK_b = K_w$.

$$K_b = \frac{[OH^-][HC_7H_5O_2]}{[C_7H_5O_2^-]} = \frac{(1.3 \times 10^5 M)(1.3 \times 10^5 M)}{1.00\ M} = 1.7 \times 10^{-10}$$

$$K_a = \frac{K_w}{K_b} = \frac{1.00 \times 10^{-14}}{1.7 \times 10^{-10}} = 5.9 \times 10^{-5}$$

PROCEDURE

You can determine the pH of the solutions investigated in this experiment using one of the three methods below. Your instructor will specify the method to be used.

Method I. pH Determination Using Indicators

Obtain about 200 mL of distilled water and boil it for 5 min to remove carbon dioxide. Carbon dioxide dissolved in water is a weak acid that will seriously affect the results of this experiment if it is not removed. For each solution to be tested, proceed as follows.

1. Rinse each of five very clean small test tubes three times with about 2 mL of boiled distilled water.

2. Rinse each test tube once with about 2 mL of the solution to be tested.

3. Place about 5 mL of the solution to be tested into each of the five test tubes.

4. Add 3 or 4 drops of one of the indicators given in Figure 18-1 to the test tubes. Use a different indicator in each test tube.

5. Record the color in each test tube and, using the data of Figure 18-1, determine and record the pH of the solution to the nearest 0.3 pH unit. (You will have to estimate the last digit.)

6. Repeat steps 1 through 5 for each solution and carefully rinse all test tubes as directed in step 1.

Method II. pH Determination Using pH Paper

pH paper is filter paper that has been impregnated with one or a mixture of indicators. The pH paper will display one of a number of colors depending on the pH of the solution to which it is exposed. pH paper is produced in two types: wide-range and short-range. Wide-range pH paper covers nearly the entire pH scale, whereas short-range pH paper may display all its colors within a pH range as narrow as 0.9 pH unit. The general procedure is to test the pH of a solution using wide-range pH paper to obtain an approximate pH. The short-range pH paper encompassing this approximate pH is then used to obtain a more exact pH.

Frequently, pH paper is sold in kits. A typical kit contains one wide-range pH paper with colors as follows.

Color:	Red	Orange	Yellow	Green	Blue
pH:	2	4	6	8	10

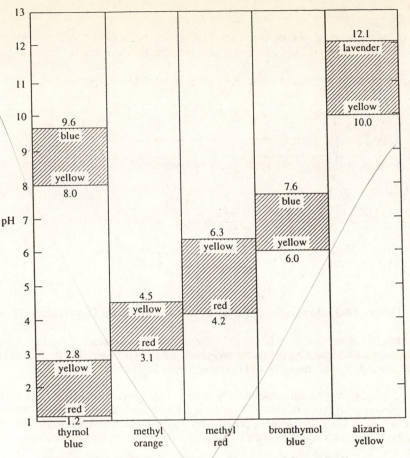

Figure 18-1. pH Color Change Ranges of Some Indicators

The kit also contains six short-range papers of the following ranges.

Range number:	1	2	3	4	5	6
pH range:	0.0–3.0	3.5–5.5	6.0–8.5	9.0–11.0	10.5–12.5	12.0–14.0

To use pH paper to determine the pH of a solution, proceed as follows:

1. Obtain 200 mL of distilled water and boil it for 5 min to remove carbon dioxide gas. Place this distilled water in your wash bottle.

2. With its range number marked in pencil, place a strip of each type of short-range pH paper on a very clean dry watch glass, as shown in Figure 18-2. Place a short strip of wide-range pH paper on another clean dry watch glass. Try to handle the pH paper as little as possible with your hands. Use paper toweling or clean forceps to hold it.

3. Rinse a clean stirring rod with boiled distilled water from your wash bottle and then with 10 to 15 drops of the solution to be tested.

4. Touch the wet stirring rod to the wide-range pH paper and record the color and approximate pH in your notebook.

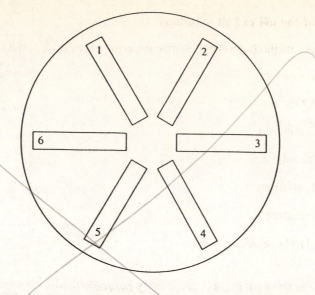

Figure 18-2. Arrangement of pH Papers on a Clean Dry Watch Glass

5. Wet the stirring rod again with 5 to 10 drops of the solution to be tested. Based on the results of step 4, select the appropriate short-range pH paper and touch the wet stirring rod to it. Record the color and pH in your notebook. Try to estimate the pH to the nearest 0.1 pH unit.

6. Repeat steps 3 through 5 for each solution to be tested. Each piece of pH paper can be reused until it is covered completely with drops of solution. After it is totally covered, it should be discarded, and the watch glass should be carefully cleaned, rinsed with distilled water, and dried. A fresh piece of pH paper then should be placed on the watch glass.

Method III. Determination of pH with a pH Meter

Although pH meters differ greatly in their method of operation, the following steps should be followed:

1. Set the temperature compensation dial of the pH meter (if there is one) to the temperature of the solution.

2. Do not leave the electrodes "high and dry." They should always be immersed in liquid except when they are being transferred from one solution to another.

3. Calibrate the pH meter (if possible) using standard pH buffer solutions. The buffers should span the range of pH that is to be investigated. Three buffers of pH = 4.0, 7.0, and 10.0 are a particularly useful set.

4. Place the solution to be investigated in the smallest container possible. A combination electrode can measure the solution in a large test tube. Two electrodes can fit easily in a 50 mL beaker. Using the smallest container possible greatly reduces the amount of solution that must be used.

5. Rinse the electrodes with CO_2–free distilled water from a wash bottle between determinations. Do not touch the electrodes at the bottom with your hands.

6. Many pH meters have a mirror behind the needle. Superimpose the needle and its mirror image when reading pH to avoid parallax.

7. Do not bump glass electrodes on the bottom or sides of the container. They are made of very thin glass at the bottom and are very fragile.

8. Record the pH in your notebook to the nearest 0.01 pH unit.

A. Determination of the pH of Salt Solutions

Use one of the three methods described to determine the pH of each of the following solutions:

1. 1.0 M NaCl solution

2. 1.0 M $NaC_2H_3O_2$ solution

3. 1.0 M Na_2SO_4 solution

4. 1.0 M $NaHSO_4$ solution

5. 1.0 M Na_2CO_3 solution

6. 1.0 M NH_4Cl solution

7. 1.0 M $NH_4C_2H_3O_2$ solution

DISPOSAL
Solutions: None of these solutions is particularly hazardous. Mix them all together in a large beaker and, using pH paper or phenolphthalein as an indicator, neutralize with 1 M HCl (the first disappearance of pink color) or 1 M NaOH (first appearance of permanent pink color). Flush the resulting salt solution down the sink with running water.

B. Determination of the Effect of Concentration on Degree of Dissociation

Use one of the three methods described previously to determine the pH of the three solutes given below at each of the following concentrations: 1.0 M, 0.10 M, and 0.010 M.

1. Acetic acid, $HC_2H_3O_2$ (aq)

2. Ammonia, NH_3(aq)

3. ~~Hydrochloric acid, HCl(aq)~~

If you are using a pH meter, do not leave the electrodes immersed in the aqueous ammonia solutions. Place the electrodes in the solution only long enough to determine the pH, turn the meter to standby, raise the electrodes, thoroughly rinse them with distilled water, and store them in pH 7 buffer or distilled water. Basic solutions can chemically attack the glass of the electrode.

DISPOSAL
Solutions: Follow the same procedure presented in Part A.

Book off limits
Re create

Report   Name _____ Section _____

DATA AND RESULTS

A. Determination of the pH of Salt Solutions

Record the pH, $[H_3O^+]$, and $[OH^-]$ of each solution.

Solution	pH		$[H_3O^+]$	$[OH^-]$
1. 1.0 M NaCl	5.4	6.12	$1 \times 10^{-6.12}$	7.59×10^{-7}
2. 1.0 M NaC₂H₃O₂	8.8	8.66	$1 \times 10^{-8.66}$	2.00×10^{-9}
3. 1.0 M Na₂SO₄	8.3	8.12	$1 \times 10^{-8.12}$	7.00×10^{-9}
4. 1.0 M NaHSO₄	1.4	1.24	$1 \times 10^{-1.24}$	5.75×10^{-2}
5. 1.0 M Na₂CO₃	11.9	11.80	$1 \times 10^{-11.80}$	1.58×10^{-12}
6. 1.0 M NH₄Cl	5.6	5.30	$1 \times 10^{-5.30}$	5.01×10^{-6}
7. 1.0 M NH₄C₂H₃O₂	7.0	7.18	$1 \times 10^{-7.18}$	1.51×10^{-7}

Salts from weak acids {2, 3}
Salt weak base {4}
Neutral {7}
Salt of weak acid + weak base

For each solute listed below, write the chemical equation for the dissociation of the salt into its ions. If the salt is acidic, write the chemical equation for the reaction of the acidic ion in water to produce hydronium ion, write the expression for the acid ionization constant, K_a and calculate the value of K_a. If the salt is basic, write the chemical equation for the reaction of the basic ion with water to produce hydroxide ion, write the expression for the base ionization constant, K_b, and calculate the value of K_a from your data.

Solute	Chemical Equations	Expression for K_b or K_a	Value of K_b or K_a
NaC₂H₃O₂	$NaC_2H_3O_2 + H_2O \rightleftharpoons H_3O^+ + NaC_2H_3O_2^-$		
Na₂SO₄	$Na_2SO_4 + H_2O \rightleftharpoons H_3O^+ +$		
NaHSO₄	$NaHSO_4 + H_2O \rightleftharpoons NaH_2SO_4 + OH^-$		
Na₂CO₃			
NH₄Cl			

Sample calc, Data Questions

Report   Name _____ Section _____

B. Effect of Concentration on Degree of Ionization

Record the pH, $[H_3O^+]$, and $[OH^-]$ of each solution that you investigated.

		1.0 M	0.10 M	0.010 M
HCl	pH			
	$[H_3O^+]$			
	$[OH^-]$			
$HC_2H_3O_2$	pH	2.36	2.70	3.08
	$[H_3O^+]$	$1.00 \times 10^{-2.36}$		
11.64	$[OH^-]$	2.29×10^{-12}		
$NH_3(aq)$	pH	11.4	10.90	9.70
	$[H_3O^+]$	$1.00 \times 10^{-11.4}$		
2.6	$[OH^-]$	2.91×10^{-3}		

Using your data, calculate the value of K_a or K_b and the degree of ionization, α, for each of the solutions you investigated.

		1.0 M	0.10 M	0.010 M
HCl	K_a			
	α			
$HC_2H_3O_2$	K_a			
	α			
$NH_3(aq)$	K_b			
	α			

Sample Calculations

Skip alpha α

Report   Name _____ Section _____

QUESTIONS

1. Which of the salts tested in the procedure were basic and which were acidic?

2. List the acidic salts in increasing order of acid strength, then list the basic salts in increasing order of base strength.

3. In Part B of the procedure, what trend (increase, decrease, or remain constant) did you observe for the values of K_a and K_b as the concentration of the acid or base decreased? Is this trend what you expected? Explain.

4. In Part B of the procedure, what trend (increase, decrease, or remain constant) did you observe for the values of the degree of ionization as the concentration of the acid or base decreased? Is this trend what you expected? Explain.

Do this one

5. When comparing solutions of equal concentration, what is the relationship between the pH of an acid solution and the acid's K_a? When comparing solutions of equal concentration, what is the relationship between the pH of a basic solution and the base's K_b?

Prelab   look future use   Name _____ Section _____

Weak Acids and Bases and Their Salts

1. The pH of a solution is measured to be 5.4. Calculate the following:

    a. $[H_3O^+]$

    b. pOH

    c. $[OH^-]$

2. In this experiment how do you determine if a salt is behaving as an acid or base?

3. Sodium fluoride, NaF, is a basic salt. Write the chemical equation that describes how the salt breaks up into its ions. Write the chemical equation that describes how one of the ions behaves as a base in water to produce OH^-.

4. The pH of a 0.10 M salt solution is found to be 8.10. Determine the K_b and the degree of ionization of the basic ion.

5. What must be done if acid or base solution is spilled on your skin?

19 Determination of the Ionization Constant of a Weak Acid

INTRODUCTION

The ionization of any weak monoprotic acid in water is most correctly represented as involving a Brønsted acid–base interaction in which water acts as the base.

$$HA + H_2O \leftrightarrows H_3O^+ + A^- \tag{19.1}$$

Since the quantity of H_2O that is used in the acid–base interaction is generally very small compared to the total quantity of water present, the concentration of water may be considered to be constant and is, therefore, not included in the expression for the ionization constant. Using H^+ to represent H_3O^+, a simpler, less-cluttered representation of the acid ionization that does not sacrifice accuracy is shown as

$$HA \overset{H_2O}{\leftrightarrows} H^+ + A^- \tag{19.2}$$

Whether the ionization is represented by equation [19.1] or [19.2] it is important to realize that, since the acid is weak, an equilibrium will exist between the molecular acid and its ionization products in solution. This equilibrium may be represented by an equilibrium constant known as an *acid ionization constant*, K_a, and defined on the basis of equation [19.1] as

$$K_a = \frac{[H_3O^+][A]}{[HA]} \tag{19.3}$$

and on the basis of equation [19.2] as

$$K_a = \frac{[H^+][A]}{[HA]}. \tag{19.4}$$

We shall use the simpler version (equation [19.4]) in this experiment.

Other expressions that are needed involve the ionization of water (the K_w expression) and the definitions of pH and of pK_a.

$$H_2O \leftrightarrows H^+ + OH^- \tag{19.5}$$

$$K_w = [H^+][OH^-] = 1.00 \times 10^{-14} \quad \text{(at 25°)} \tag{19.6}$$

$$pH = -\log[H^+] \tag{19.7}$$

$$pK_a = -\log K_a \tag{19.8}$$

Experiment 7 describes the procedure for titrating a strong acid with standard base using phenolphthalein indicator. Near the point at which equivalent quantities of acid and base have been mixed together in solution (the *equivalence point*), the phenolphthalein indicator changes color (the *end point*). The titration of a weak acid with strong base may be described as

$$HA + OH^- \leftrightarrows A^- + H_2O \tag{19.9}$$

In this experiment, using a pH meter instead of a visual indicator, we shall determine the pH at the half–equivalence point–that is, the point at which enough base has been added to neutralize exactly half the quantity of acid present. At this point, the concentration of unreacted acid, HA, is equal to the concentration of the A⁻ ion that is produced and is called the *half-equivalence point*.

$$[HA] = [A^-] \qquad\qquad [19.10]$$

Incorporation of equation [19.10] into the expression for K_a (equation [19.4]) leads to the following expression for K_a at the half-equivalence point.

$$K_a = [H^+] \qquad\qquad [19.11]$$

or $\qquad$ $pK_a = pH$ $\qquad\qquad\qquad\qquad\qquad\qquad\qquad$ [19.12]

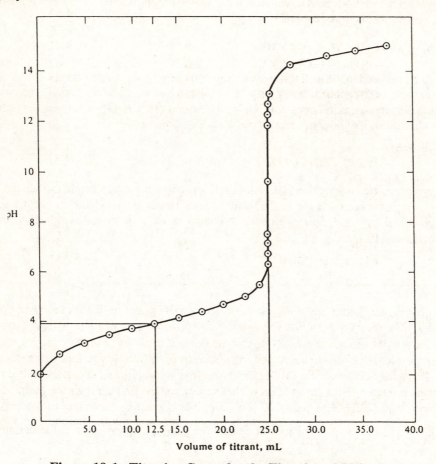

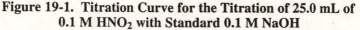

Figure 19-1. Titration Curve for the Titration of 25.0 mL of 0.1 M HNO₂ with Standard 0.1 M NaOH

Thus, if we can measure the pH at the half-equivalence point during the titration of a weak acid, we have also measured the pK_a value of the acid. In order to do this, it will be necessary to plot a *titration curve* for the weak acid–that is, to measure the pH during the titration and plot the pH as a function of mL of added base. Figure 19-1 shows a curve for the titration of the weak acid HNO₂ with standard NaOH. To obtain the data for this curve, the pH is recorded after the addition of each small increment of NaOH added (1 mL or less). The pH rises gradually until the equivalence point is approached, increases rapidly at the equivalence point with the addition of only a drop of titrant, then levels off again as excess base is added to the solution. If a vertical line is drawn to the *x* axis from the nearly vertical portion of the curve that coincides with the equivalence point, the volume of base needed to reach the equivalence point is obtained–in this example, 25.0 mL. Taking one–half this value, returning to the curve and drawing a horizontal line to the *y* axis, a pH = 3.9 at the half–equivalence point is determined. Therefore, pK_a of HNO₂ is measured to be 3.9.

In the event that the acid is much weaker than HNO_2, the change in pH in the vicinity of the equivalence point will not be as sharp and the change in slope of the titration curve will be more gradual. In order to pinpoint the equivalence point under these circumstances, it is necessary to locate the point of inflection in the curve–specifically, that point at which the curve changes direction. This point is the equivalence point.

Table 19-1. Properties of Some Weak Acids

Acid	Formula	Molar Mass, g/mol	pK_{a1}	pK_{a2}
Acetic	CH_3COOH	60.05	4.75	
Acetylsalicylic	$CH_3COOC_6H_4COOH$	180.2	3.49	
Benzoic	C_6H_5COOH	122.12	4.19	
Butyric	$CH_3(CH_2)_2COOH$	88.1	4.82	
Carbonic	H_2CO_3	62.0	6.34	10.33
trans-Cinnamic	$C_6H_5CH=CHCOOH$	148.15	4.44	
Fumaric	$HOOCCH=CHCOOH$	116.1	3.02	4.39
Malonic	$HOOCCH_2COOH$	104.06	2.83	5.69
Mandelic	$C_6H_5CH(OH)COOH$	152.14	3.40	
Oxalic 2-hydrate	$H_2C_2O_4 \cdot 2\ H_2O$	126.07	1.27	4.27
Phthalic	$C_6H_4(COOH)_2$	166.1	2.89	5.41
Potassium hydrogen phthalate	$HOOCC_6H_4COOK$	204.23	5.41	
Sodium bisulfate	$NaHSO_4 \cdot H_2O$	138	1.92	
Sodium bisulfite	$NaHSO_3$	104.06	7.20	
Sodium dihydrogen phosphate	$NaH_2PO_4 \cdot H_2O$	138	7.21	
Sulfurous	H_2SO_3	82.0	1.89	7.20
Valeric	$CH_3(CH_2)_3COOH$	102.13	4.82	

If your unknown acid is diprotic, two inflection points will appear on the titration curve (Figure 19-2). In this event, *both* pK_{a1} and pK_{a2} must be determined. The consideration of pK_{a1} and pK_{a2} along with its molar mass should identify the unknown acid. Note that the titration curve for a diprotic acid contains two inflection points—two equivalence points. In the region from the start of the titration to the first equivalence point (point A on Figure 19-2), the first proton is reacting according to reaction [19.13]. Thus, the half-equivalence point of this first titration is midway between the start of the titration and the first equivalence point. In the region from the first to the second equivalence point, the second proton is reacting according to reaction [19.14]–thus, the half-equivalence point for pK_{a2} is midway between the two equivalence points. In most cases the first inflection will not be as large as that in Figure 19-2.

$$H_2SO_3(aq) + OH^-(aq) \leftrightarrows HSO_3^-(aq) + H_2O(l) \qquad [19.13]$$

$$HSO_3^-(aq) + OH^-(aq) \leftrightarrows SO_3^{2-}(aq) + H_2O(l) \qquad [19.14]$$

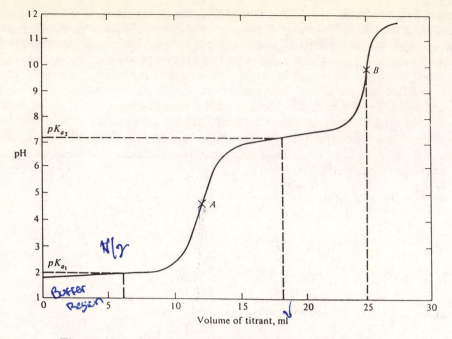

Figure 19-2. Curve Showing the Titration of 0.500 M H₂SO₃ Solution with 0.100 M NaOH

SAFETY PRECAUTIONS

Review the safety rules on pages 1 and 2.

Acids and bases are harmful to skin and eyes. Wear eye protection. If any acid or base spills on skin or eyes, rinse thoroughly with running water for at least 5 min.

PROCEDURE

A. Determination of Sample Size of Unknown Acid

It will first be necessary to determine the appropriate sample size, one that will require about 25 mL of standard base for titration. In order to do this, 0.2 g of unknown acid is titrated to the phenolphthalein end point. Weigh about 0.2 g of unknown acid into a clean Erlenmeyer flask and record the mass to ±0.001 g. Add enough distilled water to the flask to dissolve the acid (about 50 mL). Add 2 to 3 drops of phenolphthalein indicator and titrate with the standard NaOH solution to a pink end point. Before titrating, be sure to rinse the buret properly according to the procedure in Experiment 7.

B. Determination of pK_a of the Unknown Acid

(See Experiment 18, Method III, for the proper use of a pH meter.) From the titration data in Part A, calculate the mass of unknown acid that will require 25.00 mL of standard NaOH for complete titration.

$$\text{Grams of unknown acid needed} = \frac{\text{Mass of unknown acid, g (PartA)}}{\text{mL of NaOH used (PartA)}} \times 25.00 \text{ mL}$$

Weigh the calculated quantity of unknown acid to the nearest mg (±0.001 g) into a clean 250 mL beaker. Add 50 mL of distilled water to dissolve all the acid in the beaker. Obtain a calibrated pH meter and immerse the electrodes into the acid solution in the beaker along with a magnetic stirring bar. The tip of the glass pH electrode is very thin and fragile. *Therefore, ensure that sufficient space is allowed between the stirring bar and the tips of the electrodes so that they will not come into contact.* While stirring slowly with the magnetic stirrer, titrate the acid with standard NaOH, recording the pH and mL NaOH after each addition of titrant. The NaOH solution can be added in 1 mL increments until the pH changes by more than 0.2-0.3 pH unit with each mL. More than a few mL of titrant will have been added before the pH changes are greater than 0.2-0.3 pH units. At this point, reduce the size of the increment to 0.1-0.2 mL, eventually adding the NaOH solution dropwise near the equivalence point. After the equivalence point has been passed, the increment of titrant can be readjusted to about 1 mL. Continue to add the NaOH solution until the change in pH with each mL of base added is approximately what it was at the beginning of the titration. This will ordinarily occur before a pH of 11.0 has been reached. Record all your data in your laboratory notebook. You may even plot your pH versus mL NaOH data as you titrate (see Figure 19-2).

DISPOSAL

Solutions: None of these solutions is particularly hazardous. Mix them all together in a large beaker and, using pH paper or phenolphthalein as an indicator, neutralize with 1 M HCl (the first disappearance of pink color). Flush the resulting salt solution in the sink with running water.

Due Tues 5 3/23

Data Can use data table

Sample calc

Prelab   Name _____ Section _____

Determination of the Ionization Constant of a Weak Acid

    1. What is the precaution that you must take regarding the use of the stirring bar?

    2. What is the difference between a monoprotic and diprotic acid? How would their titration curves differ?

    3. What is the difference between an equivalence point and an end point?

    4. A 0.200 gram sample of a weak monoportic acid required 29.41 mL of 0.100 M NaOH for neutralization. Calculate the molar mass of the acid.

    5. At the beginning of the titration, how much titrant should be added between pH readings? As you approach the equivalence point, how much titrant should be added between pH readings?

Report

DATA

215-493
6138
.00673

Unknown number _____#1_____

Mass of unknown acid, Part A _____.202_____ g

Concentration of standard NaOH solution _____0.0504_____ mol/L

Volume of NaOH solution, Part A _____29.00_____ mL

Moles of NaOH solution, Part A(= moles of H+ from acid) _____0.00146 mols_____

Molar mass of acid = $\dfrac{\text{Mass of acid used}}{\text{Mol H}^+ \text{ ion produced}}$

(× 2 if the acid is diprotic), Part A _____138.36_____ g/mol

Mass of unknown acid, Part B _____.172_____ g

Volume of NaOH solution needed to
reach the equivalence point, Part B _____24.2_____ mL

Moles of NaOH, Part B (= moles of H+ from acid) _____0.00122 Moles_____

Molar mass of acid = $\dfrac{\text{Mass of acid used}}{\text{Mol H}^+ \text{ ion produced}}$

(× 2 if the acid is diprotic), Part B _____140.98_____ g/mol

pK_a of unknown acid _____

K_a of unknown acid _____

Name of unknown acid (use Table 19-1) _____

Sample Calculations

Lexi 23693

$0.0504 \dfrac{mol}{L} \times 0.029 L =$

$\dfrac{.202 \times 25.00}{29.00} = .178 g$

$\dfrac{.202}{.00302} = 66.9 \text{ g/mol}$

$\dfrac{.172}{0.0022}$

0.01202

Report   Name _____ Section _____

TITRATION CURVE DATA FOR PART B

pH	Volume of Titrant (mL)	pH	Volume of Titrant (mL)	pH	Volume of Titrant (mL)
1.8	0	2.1	15.2	11.8	31.1
1.6	.90	2.2	16.6	11.8	32.0
1.6	2.00	2.3	17.9	11.9	33.2
1.7	3.00	2.4	19.1	11.9	34.2
1.7	4.10	2.5	20.1		
1.7	5.00	2.6	21.0		
1.8	6.00	2.7	22.0		
1.8	7.00	3.0	23.2		
1.8	8.10	3.5	24.1		
1.8	9.10	10.4	25.2		
1.8	10.0	11.1	26.2		
1.9	11.1	11.4	27.3		
1.9	12.2	11.5	28.2		
2.0	13.3	11.6	29.1		
2.0	14.2	11.7	30.2		

Alison
Butler

Report   Name _____ Section _____

QUESTIONS

1. Plot the pH of the solution on the ordinate and the volume of standard base on the abscissa on linear graph paper. Refer to the section in the Introduction describing good graphing technique.

2. Calculate the number of moles of NaOH solution needed to reach the equivalence point.

3. Calculate the molar mass of the unknown acid. Two values for the molar mass can be calculated if the data from both Parts A and B are used.

sample
calcs
don't
write up

4. From your titration curve, determine the pK_a and K_a of your unknown acid according to the procedure described in the Introduction. Record their values on the report sheet.

5. Identify the unknown acid by comparing its experimental pK_a and molar mass with those listed in Table 19-1. Record your unknown acid on the report sheet.

6. How would the pK_a of the unknown acid be affected (higher, lower, or no change) if the following errors occurred? Briefly explain why in each case.

    a. The pH meter was incorrectly calibrated to read lower than the actual pH.

    b. During the titration several drops of NaOH missed the reaction beaker and fell onto the bench top.

    c. In Part B of the procedure the acid was dissolved in 75 mL of distilled water rather than the 50 mL called for in the procedure.

7. How would the molar mass of the unknown acid be affected (higher, lower, or no change) if the following errors occurred? Briefly explain why in each case.

    a. The pH meter was incorrectly calibrated to read lower than the actual pH.

    b. During the titration, several drops of NaOH missed the reaction beaker and fell onto the bench top.

    c. In Part B of the procedure the acid was dissolved in 75 mL of distilled water rather than the 50 mL called for in the procedure.

Data   Table of
            PH & Vol acid
                + base added
                    (both buffers)

2 Results initial pH of ⎫ measure
            both buffers   ⎭ & calc

            buffer capacity
            acid x base — both buffers

3 Graph

4. Sample calc
        both initial pH calc
        buffer cap (1 on 1 ✗

Data Table
One Graph w/2 sets or data
Sample calc   (initial pH of both buffers)
              (Henderson eq)
buffer capacity calc

No Dis
No Quest.

alif lower |||||||
            (7)

20 Investigation of Buffer Systems

INTRODUCTION

The purpose of this experiment is to study the effect of adding small amounts of strong acids and bases to several buffer systems and to determine how effectively each system resists large changes in pH.

An important characteristic of weak acids and bases is their ability to form buffer systems. A buffer is a solution whose pH does not change significantly when small amounts of a strong base or acid are added. This definition is somewhat inexact and the words "significantly" and "small amounts" are interpreted differently depending on the context. For example, the pH of normal human blood ranges from 7.35 to 7.45, so a pH change of 0.10 unit would be significant. On the other hand, the indicator phenolphthalein changes color in the pH range of 8 to 10, so here a pH change would have to be in the order of 1.0 unit to be significant.

The action of a buffer is caused by the presence of *both* the weak acid and its anion (or the weak base and its cation)–that is, a conjugate acid–base pair. Consider a sodium acetate–acetic acid buffer system in which there are approximately equimolar amounts of acetic acid ($HC_2H_3O_2$) and its conjugate base, acetate ion ($C_2H_3O_2^-$). The addition of a strong acid adds hydronium ion (H_3O^+) to the solution. The acetate ion (the base form) in the buffer reacts with the hydronium ion to neutralize it.

$$H_3O^+(aq) + C_2H_3O_2^-(aq) \leftrightarrows H_2O + HC_2H_3O_2(aq) \qquad [20.1]$$

In like fashion, the addition of a strong base adds hydroxide ion (OH^-) to the solution. Then the acetic acid reacts with this ion to neutralize it.

$$OH^-(aq) + HC_2H_3O_2 (aq) \leftrightarrows H_2O + C_2H_3O_2^-(aq) \qquad [20.2]$$

There are two features common to reactions [20.1] and [20.2]. First, the ion characteristic of a strong acid (H_3O^+) or a strong base (OH^-) is removed by reaction with one of the components of the buffer, either the conjugate base or the conjugate acid. Second, water and the other component (the conjugate acid or the conjugate base) of the buffer are produced as products. Since both OH^- and H_3O^+ are removed by reaction with the components of the buffer and are replaced by weak products, the pH of the solution will not change by a large amount.

Eventually a buffer will cease to retard the change of pH when all of the reacting component (either the conjugate acid or the conjugate base) is used up by the stronger acid or base. For example, if a buffer solution contains 0.010 mol $C_2H_3O_2^-$, the addition of 100 mL of 0.10 M HCl (which contains 0.010 mol H_3O^+) will consume all the acetate ion by reaction [20.1]. Further addition of strong acid will cause a rapid drop in pH, since there is no base (except H_2O) present to react with the added H_3O^+. The amount of strong acid or base that can be added to a given volume of a buffer system without a significant change in pH (±1.0 unit) is known as the *buffer capacity*.

The pH of a buffer solution can be determined exactly by measurement or approximately by calculation. The equilibrium that exists for the weak acid HA in water is described by the following equation.

$$HA(aq) + H_2O \leftrightarrows H_3O^+(aq) + A^-(aq) \qquad [20.3]$$

For this reaction, the acid ionization constant expression is given in equation [20.4].

$$K_a = \frac{[H_3O^+][A^-]}{[HA]} \qquad\qquad [20.4]$$

Rearrangement of equation [20.4] yields the *Henderson-Hasselbalch equation* [20.5].

$$\log K_a = \log [H_3O^+] + \log \frac{[A^-]}{[HA]}$$

$$-\log [H_3O^+] = -\log K_a + \log \frac{[A^-]}{[HA]}$$

$$pH = pK_a + \log \frac{[A^-]}{[HA]} \qquad\qquad [20.5]$$

Consideration of equation [20.5] shows that the pH of a buffer depends on both the pK_a of the acid and the ratio of the concentration of anion to that of un-ionized acid. In order to produce a buffer with a pH value different from the value of the pK_a of the acid, one only needs to adjust this ratio appropriately. However, if the ratio differs significantly from unity, the buffer's capacity toward a strong acid will be different from its capacity toward strong base, and its buffer capacity is diminished. Consequently, a buffer is usually formulated using an acid–base system whose pK_a approximates the pH desired.

The Henderson-Hasselbalch equation allows a more precise definition of buffer capacity. When the ratio $[A^-]/[HA]$ in equation [20.5] is less than 0.10 or greater than 10.0, the useful buffer capacity has been exceeded. Thus, the useful pH range of a buffer extends one pH unit above or below the pK_a of the weak acid used. For example, the useful pH range of the acetic acid-acetate ion buffer system is from 3.75 to 5.75, since acetic acid has a pK_a of 4.75.

SAFETY PRECAUTIONS

Review the safety rules on pages 1 and 2.

Acids and bases are harmful to skin and eyes. Wear eye protection. If you spill any on yourself, immediately wash thoroughly with soap and running water for at least 5 min. Tell your instructor about the spill.

Wear goggles at all times. If you do not and get acid or base in your eye, immediately wash with large volumes of running water, and get emergency medical attention. Do not delay if you want to save your vision.

PROCEDURE

A. Formulation of Buffers

For one of the acid solutions given below, prepare a buffer solution as described in steps 1 through 3. Use the acid your instructor directs.

Acid	K_a	pK_a
Phthalic acid, $C_6H_4(COOH)_2$	1.3×10^{-3}	2.89
Formic acid, HCOOH	1.8×10^{-4}	3.74
Acetic acid, CH_3COOH $HC_2H_3O_2$	1.8×10^{-5}	4.74
Potassium hydrogen phthalate (KHP), $C_6H_4COOHCOOK$	3.9×10^{-6}	5.41
Potassium dihydrogen phosphate, KH_2PO_4	6.2×10^{-8}	7.21

Graduated Cylinders

Don't boil Water

1. If boiled distilled water is not available in the laboratory, place 200 mL of distilled water in a beaker, heat to boiling for 5 min to drive off dissolved carbon dioxide, and allow to cool. Carbon dioxide is a weak acid that may significantly affect the results of this experiment if it is present in solution.

2. Pipet 20.0 mL of the 0.100 M acid solution into a 100 mL beaker and add 20.0 mL of 0.100 M NaOH with a pipet. This produces 40.0 mL of 0.0500 M solution of the acid's anion, $[A^-] = 0.0500$ M.

weak acid

3. Pipet 20.0 mL of the 0.100 M acid solution into another 100 mL beaker and add 20.0 mL of cooled distilled water with a pipet. This produces 40.0 mL of 0.0500 M solution of the undissociated acid, $[HA] = 0.0500$ M.

4. Mix the two solutions together. This produces 80.0 mL of solution in which $[HA] = 0.0250$ M and $[A^-] = 0.0250$ M. This solution will be referred to as buffer *a*.

5. Pipet 40.0 mL of 0.100 M acid solution into a 100 mL beaker. Add 10.0 mL of boiled distilled water (cooled). Add 30.0 mL of 0.100 M NaOH solution from a buret. This will produce 80.0 mL of solution in which the concentration of the acid's anion is about three times that of the undissociated acid. This solution will be referred to as buffer *b*.

No Calcs today

B. Measurement of pH and Determination of Buffer Capacity

Note: In what follows, you may find it quite difficult to change the pH by exactly 0.40 pH unit each time. Do the best you can, but be sure to record the exact pH obtained and the exact volume of acid or base added. You should be able to estimate both pH and milliliter volume to ± 0.01 units. (See Experiment 18, Method III, for the proper use of a pH meter.)

1. Measure 25.0 mL of buffer solution *a* into a 50 mL [*250 mL*] beaker, measure the pH with a pH meter, and record the pH.

2. Add 0.10 M HCl solution dropwise from a buret until the pH changes by 0.40 pH unit. Stir constantly while the strong acid is being added. Record the pH and the volume of HCl solution added.

3. Continue to add HCl solution dropwise. Stop when the pH has changed a total of 0.80, 1.20, 1.60, and 2.00 pH units. Record the *total* volume of HCl solution added and the pH at each point.

4. Repeat steps 1 through 3 with 25.0 mL of buffer *b*.

5. Measure 25.0 mL of cool boiled distilled water into a 50 mL beaker, measure the pH with a pH meter, and record the pH. Add 5-drop increments 0.10 M HCl solution from a buret recording the pH and the *total* volume of strong acid added after each increment until the pH changes by 5 pH units. Stir constantly while the strong acid is being added.

6. Transfer 25.0 mL of the remainder of buffer *a* to another 50 mL beaker.

7. Add 0.10 M NaOH solution dropwise from a buret until the pH changes by 0.40 pH unit. Stir constantly while the strong base is being added. Record the pH and the volume of NaOH solution added.

8. Continue to add NaOH solution dropwise. Stop when the pH has changed a total of 0.80, 1.20, 1.60, 2.00, and 3.00 pH units. Record the *total* volume of NaOH solution added and the pH at each point.

9. Repeat steps 6 through 8 with 25.0 mL of buffer *b*.

10. Measure 25.0 mL of cool, boiled distilled water into a 50 mL beaker, measure the pH with a pH meter, and record the pH. Add 5-drop increments 0.10 M NaOH solution from a buret recording the pH and

the *total* volume of strong base added after each increment until the pH changes by 5 pH units.   Stir constantly while the strong base is being added.

DISPOSAL

Solutions: None of these solutions is particularly hazardous.   Mix them all together in a large beaker and, using pH paper or phenolphthalein as an indicator, neutralize with 1 M HCl (the first disappearance of pink color) or 1 M NaOH (first appearance of permanent pink color).   Flush the resulting salt solution down the sink with running water.

C. Reporting Data

For each buffer, make a plot of pH versus volume of strong acid and strong base.   All three plots should be done on the same graph.   Use different colors or different symbols for data points to distinguish between the three plots.   Plot the pH of each buffer at a volume of zero midway on the *x*-axis as shown in Figure 20-1.   Plot the volumes of strong base added and the corresponding pH values in the positive direction on the *x*-axis beginning at the point labeled *buffer*.   Plot the volumes of strong acid added and the corresponding pH values in the negative direction along the *x*-axis beginning at the point labeled *buffer*.   Choose a scale for the *y*-axis that will range from the lowest pH value measured to the highest pH measured for the three buffers.   Choose a scale for the *x*-axis that will range from the largest amount of acid added (to the left of the point labeled *buffer*) to the largest volume of base added (to the right of the point labeled *buffer*).   Make your graph large enough so that the three plots are not crowded.

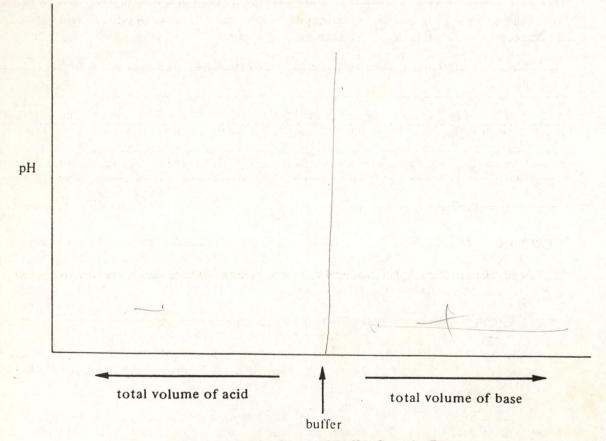

Figure 20-1. Axes for the Graph of Buffer Capacity Data

Report

Name _____ Section _____

DATA

Record the *total* volumes, in milliliters, of 0.100 M NaOH and 0.100 M HCl added and the observed pH for each buffer and for water.

Buffer *a*		Buffer *b*		Distilled Water	
Initial pH 7.3		8.3			
Added Acid, mL	pH	Added Acid, mL	pH	Added Acid, mL	pH
2.8 mL	6.9	1.3	7.7		
2.8 mL	6.5	1.5	7.3		
1.4 mL	6.1	2.5	6.9		
1.2 mL	5.7	2.3	6.5		
0.7 mL	4.2	2.3	6.1		

Added Base, mL	pH	Added Base, mL	pH	Added Base, mL	pH
2.0 mL	7.7	.6	9.6		
0.9	8.2	.2	10.0		
1.0	8.6	3.3	10.4		
.4	9.9	0.6	10.8		
.4	10.4	-1.1	11.2		

RESULTS

Include the three plots of pH versus total volume of strong acid and of strong base with your report.

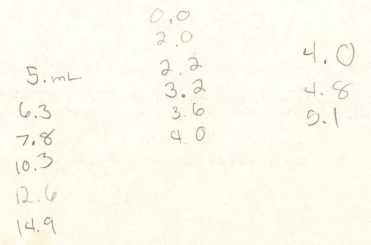

0.0
2.0
2.2
3.2
3.6
4 0

4.0
4.8
5.1

5. mL
6.3
7.8
10.3
12.6
14.9

Report   Name _____ Section _____

QUESTIONS

1. Which of the three solutions (buffer *a*, buffer *b*, or water) had the highest buffer capacity for the addition of a strong acid? What experimental evidence do you have to support this?

2. Which of the three solutions (buffer *a*, buffer *b*, or water) had the highest buffer capacity for the addition of a strong base? What experimental evidence do you have to support this?

3. Which of the three solutions (buffer *a*, buffer *b*, or water) had the smallest buffer capacity for the addition of both acid and base? What experimental evidence do you have to support this?

4. Use the Henderson-Hasselbalch equation and the method of preparation to calculate the predicted pH of the buffer *a* and *b* which you prepared. Compare the predicted pH to the pH which you measured.

5. Buffer capacity is normally defined as the millimoles of strong acid or base required to change the pH of 1 liter of buffer by ± 1 pH unit. Use your graph to estimate the buffer capacity of water, buffer *a*, and buffer *b*.

Prelab   Name _____ Section _____

Investigation of Buffer Systems

1. Buffers are made from weak conjugate acid-base pairs. In Part A, step 4, of the procedure a weak base, A^- (prepared in step 2), is mixed with a weak acid, HA (prepared in step 3), to make a buffer. In Part A, step 5, of the procedure a buffer is prepared by adding 30.0 mL of strong base NaOH to 40.0 mL of weak acid. Explain how the procedure in step 5 is able to produce a buffer. Use chemical equations (HA and A^-) to support your answer.

2. Calculate the concentration of both HA and A^- in the buffer produced in Part A, step 5.

3. Why is it necessary to boil the distilled water before using it in this experiment?

4. Write chemical formulas for the conjugate base for each substance in the table in Part A of the procedure. Note that two of the substances are salts. You will need to consider the ions produced from these two substances.

5. Which substance in the table of Part A of the procedure would be best suited to make a buffer of pH 3.00.

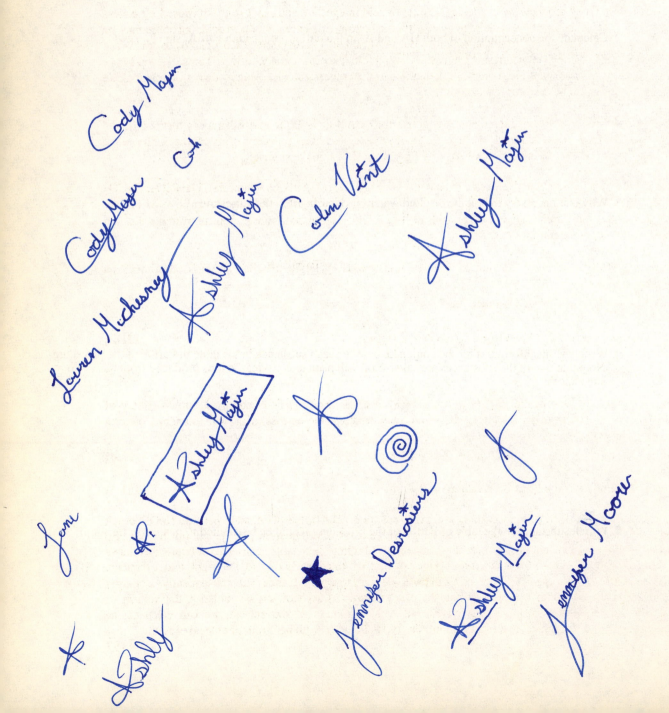

22 Determination of a Solubility Product Constant

INTRODUCTION

The purpose of this experiment is to determine the solubility product constant, K_{sp}, of the sparingly soluble salt silver acetate, $AgC_2H_3O_2$. This experiment will also investigate the independence of the solubility product from concentration by forming silver acetate from three solutions of different concentrations while the dependence of the solubility product on temperature is verified by forming the precipitate at three different temperatures.

Many salts are sparingly soluble in water. These salts are frequently described as insoluble rather than as sparingly soluble. However, insoluble implies that none of the salt dissolves in solution, which is definitely not the case. For example, a saturated solution of the "insoluble" salt calcium carbonate, $CaCO_3$, contains 0.0070 g of the salt in a liter of solution. Although this does not seem to be a large mass of compound, it represents more than 8×10^{19} ions.

In its saturated solutions, a sparingly soluble salt is in equilibrium with its dissolved constituent ions. In general, for a salt having a formula M_aX_b, the solubility equilibrium may be expressed as follows.

$$M_aX_b(s) \leftrightarrows a\, M^{b+}(aq) + b\, X^{a-}(aq) \qquad [22.1]$$

The equilibrium constant for this reaction is called the solubility product constant, K_{sp}, shown as

$$K_{sp} = [M^{b+}]^a[X^{a-}]^b \qquad [22.2]$$

For example, silver chloride's solubility equilibrium in water is represented by

$$AgCl(s) \leftrightarrows Ag^+(aq) + Cl^-(aq) \qquad [22.3]$$

The solubility product constant, K_{sp}, is given by

$$K_{sp} = [Ag^+][Cl^-] = 1.7 \times 10^{-10} \qquad [22.4]$$

For a given salt, the value of the solubility product constant depends only on temperature. K_{sp} does not depend on the relative concentrations of the two ions. For example, several maximum concentration combinations of silver and chloride ions that can exist in solution before precipitation occurs are listed below.

$[Ag^+]$, M	1.3×10^{-5}	1.7×10^{-4}	2.1×10^{-7}	6.0×10^{-5}
$[Cl^-]$, M	1.3×10^{-5}	1.0×10^{-6}	8.1×10^{-4}	2.8×10^{-6}

In this experiment, solutions of sodium acetate and silver nitrate are combined to form a solution in which the ion product of silver acetate exceeds its solubility product. Consequently, solid silver acetate precipitates, leaving behind a solution in which the concentrations of silver ion and acetate ion obey the solubility product expression, shown below as

$$K_{sp} = [Ag^+][C_2H_3O_2^-] \qquad [22.5]$$

The saturated solution is filtered to remove the precipitate and the concentration of silver ion [Ag^+] is determined by titration with a standard potassium thiocyanate, KCNS, solution. The thiocyanate ion reacts with silver ion to form the very sparingly soluble white precipitate silver thiocyanate, AgSCN, as follows:

$$Ag^+(aq) + SCN^-(aq) \rightleftharpoons AgSCN(s) \qquad [22.6]$$

The end point of the titration is indicated by the production of a bright red color in solution by the reaction between Fe^{3+} ions and SCN^- ions.

$$Fe^{3+}(aq) + SCN^-(aq) \rightleftharpoons FeSCN^{2+}(aq) \qquad [22.7]$$

Until nearly all the silver ions precipitate as AgSCN, there is an insufficient concentration of thiocyanate ion to permanently produce the bright red color. Once precipitation of AgSCN is complete, the color of $FeSCN^{2+}$ remains permanent with the addition of one drop of KSCN solution in excess.

Once the concentration of silver ion is determined, the acetate ion concentration can be calculated since the numbers of moles of both ions that precipitate are equal. Of course, it is necessary to know the initial concentrations of silver ion and acetate ion in the solution before the reaction begins to occur.

SAFETY PRECAUTIONS

Review the safety rules on pages 1 and 2.

Wear goggles to protect your eyes.

To avoid ingesting solution when pipetting, use a suction device, never your mouth.

Ag^+ solutions will stain your hands dark. This unsightly stain of elemental silver will wear off in a few weeks. Use good technique, work carefully, and clean up all spills immediately .

PROCEDURE

Part I.   DEPENDENCE ON CONCENTRATION

A. Preparation of Saturated Silver Acetate Solutions

1. The concentrations of all solutions must be known to three significant figures. All volumes must be measured to the nearest 0.1 mL. Make sure that all glassware is clean and dry. Dirty glassware will introduce contamination, and wet glassware will dilute the solutions being used.
2. Measure into 250 mL Erlenmeyer flasks the volumes of reagents indicated in Table 22-1 to make the three saturated solutions. The volumes should be measured with a pipet. It may be impractical to use dry pipets. Thus, each pipet should be rinsed twice with distilled water and twice with the solution being measured. Use small volumes (5 mL or less) to rinse the pipets. Label these solutions *a*, *b*, and *c*, as shown in Table 22-1.

Table 22-1. Volumes of Stock Solution

	Volume to Be Used, mL	
Solution	0.300 M $NaC_2H_3O_2$	0.250 M $AgNO_3$
a	20.0	35.0
b	30.0	25.0
c	25.0	30.0

3. The solution in each flask should be stirred or shaken until precipitation begins and swirling should be done every 5 minutes for at least 30 min to ensure that equilibrium is established.
4. If you are going to perform Part II, record the temperature of the solution in flask *c* before filtering. Make sure that the thermometer is scrupulously clean and dry.
5. Filter each solution through *dry* filter paper in a clean, *dry* funnel into a clean, *dry* 250 mL beaker.

B. Titration of Saturated Silver Acetate Solutions

1. Use a clean pipet. It should be dry if possible, but at least it should be rinsed with two 5-mL portions of the saturated solution. Transfer 20.0 mL of one saturated solution into a clean container of at least 150 mL volume.
2. Add 2 mL of the indicator [a saturated solution of $(NH_4)Fe(SO_4)_2$ in 1 M HNO_3].
3. Titrate the solution with standard KSCN solution, stirring constantly. Note that red $FeSCN^{2+}$ forms when the titrant is added, but the color disappears as the solution is stirred. Continue to add KSCN solution slowly with stirring and eventually 1 drop at a time until the red color persists. Record the initial and final volumes of titrant in the buret.
4. Repeat steps 1 through 3 for the remaining two saturated solutions.
5. If time permits, perform a second titration of each saturated solution. A second determination will significantly enhance the precision of your results.

Part II. DEPENDENCE ON TEMPERATURE

A. Formulation of Saturated Silver Acetate Solutions

1. Obtain three clean, dry 125 mL Erlenmeyer flasks. Label these flasks, *c*, *d*, and *e*. If you have done Part I, you need only two flasks, labeled *d* and *e*.
2. Following the procedure in Part I.A, steps 1 and 2, pipet 25.00 mL of 0.30 M $NaC_2H_3O_2$ and 30.00 mL of 0.25 M $AgNO_3$ into each flask.
3. Stir or shake each flask until precipitation begins. Stopper each flask.
4. If you have not done Part I, treat flask **c** as described in Part I.A, steps 3 through 5.
5. Place 150 mL of crushed ice in a 400 mL beaker. Place flask *d* on top of the ice and pack crushed ice around it. Add tap water to the beaker up to the 200 mL mark. Agitate the flask periodically for 30 min to ensure that equilibrium is achieved.
6. Place 150 mL of distilled water in another 400 mL beaker and place it on a ring stand using a wire gauze support with a ceramic center. Place flask *e* in the water and warm the water to 45°C using a Bunsen burner. Maintain the water as close to 45°C as possible by intermittent use of the Bunsen burner. Agitate the flask periodically for 30 min to ensure that equilibrium is achieved.
7. Record the temperature of solution *d*. Then remove the flask from the beaker and filter the solution through dry filter paper in a clean, dry funnel into a clean, dry 250 mL beaker. You will obtain better results if both the funnel and the filter paper have been stored for at least 30 min in a refrigerator.
8. Record the temperature of solution *e*. Pipet 25.00 mL of distilled water into a clean, dry 250 mL beaker. The distilled water is added to prevent precipitation as the solution cools. Remove flask *e* from the 500 mL beaker. Filter the solution through dry paper in a clean, dry funnel into the 250 mL beaker. You will obtain better results if both the funnel and the filter paper have been stored for at least 30 min in a drying oven at 50 to 60°C. After the filtration is complete, thoroughly stir the solution in the beaker to make sure that the filtrate is completely mixed with the distilled water.

B. Titration of Saturated Silver Acetate Solutions

1. Follow exactly the same procedure described in Part I.B, steps 1 through 5.
2. Note that the computations involving the data from flask *e* are performed somewhat differently from those of the other flasks.

DISPOSAL

Solid silver acetate wastes: *Dispose* solid and filter paper in a waste bottle labeled for *silver acetate*.

Solid silver thiocyanate wastes: *Dispose* solid and filter paper in a waste bottle labeled for *silver thiocyanate*.

Titration solutions: *Dispose* in a waste bottle labeled *heavy metals—solutions*.

CALCULATIONS (FOR BOTH PARTS I AND II)

1. Assume that the total volume of each saturated solution is 55.00 mL.
2. Determine the initial (before precipitation) numbers of millimoles of silver ions and acetate ions in the three solutions by multiplying the volumes of the stock solutions used by their concentrations.
3. Determine the millimoles of silver ion in each saturated solution as follows. The number of millimoles of silver ion in 20.00 mL of solution is the volume of titrant multiplied by the molarity of the titrant. This result multiplied by the ratio 55:20 is the number of millimoles of silver ion in the saturated solution. (Because solution *e* of Part II was diluted with 25.00 mL of distilled water, its total volume was 80.00 mL rather than 55.00 mL. Consequently, the millimoles of silver ion should be multiplied by the ratio 80:20 to obtain the number of millimoles of silver ion in the saturated solution. All other calculations should be performed as described in the other steps of this section.)
4. The millimoles of silver ion and also of acetate ion in the precipitate are determined by subtracting the number of millimoles of silver ion in the saturated solution (computed in step 3) from the number of millimoles of silver ion initially present, computed in step 2.
5. The millimoles of acetate ion in the saturated solution is the result of subtracting the number of millimoles of acetate in the precipitate (step 4) from the number of initial millimoles of acetate present (step 2).
6. The concentration of the two ions is the number of millimoles in the saturated solution (step 3 and step 5) divided by 55.00 mL, the volume of the saturated solution.

Report   Name _____ Section _____

DATA

Part I. Dependence on Concentration

1. Concentration of supplied solutions.

Solute	$AgNO_3$	$NaC_2H_3O_2$	KSCN
Concentration, M	_____	_____	_____

2. Millimoles of ions in the three solutions initially.

	Soln. *a*	Soln. *b*	Soln. *c*
Ag^+	_____	_____	_____
$C_2H_3O_2^-$	_____	_____	_____

3. Volume of titrant used.

	Soln. *a*	Soln. *b*	Soln. *c*
Trial 1	_____	_____	_____
Trial 2	_____	_____	_____

4. Millimoles of Ag^+ ion in solution and in precipitate.

		Soln. *a*	Soln. *b*	Soln. *c*
Solution:	Trial 1	_____	_____	_____
	Trial 2	_____	_____	_____
Precipitate:	Trial 1	_____	_____	_____
	Trial 2	_____	_____	_____

5. Millimoles of acetate ion, $C_2H_3O_2^-$, in solution.

	Soln. *a*	Soln. *b*	Soln. *c*
Trial 1	_____	_____	_____
Trial 2	_____	_____	_____

Report   Name _____ Section _____

6.   Concentration of ions and solubility products.

		Soln. *a*	**Soln. *b***	**Soln. *c***
$[Ag^+]$:	Trial 1	_____	_____	_____
	Trial 2	_____	_____	_____
$[C_2H_3O_2^-]$:	Trial 1	_____	_____	_____
	Trial 2	_____	_____	_____
K_{sp}:	Trial 1	_____	_____	_____
	Trial 2	_____	_____	_____
	Average K_{sp}	_____	_____	_____

7.   The accepted value of the solubility of silver acetate (FW = 166.9 g/mol) at 20°C is 1.20 g/100 mL. Use this value to calculate K_{sp} and the % error in each of your measured values. (In different reference works K_{sp} values often differ by a factor of 10 or more, depending on methods of analysis used.)

K_{sp} calculated from accepted solubility = _____

	Soln. *a*	**Soln. *b***	**Soln. *c***
Error in K_{sp}:	_____ %	_____ %	_____ %

8.   Is your error within experimental error or due to a personal error? Please explain.

Sample Calculations: Part I

Report Name _____ Section _____

Part II. Dependence on Temperature

1. Concentration of supplied solutions.

Solute	$AgNO_3$	$NaC_2H_3O_2$	KSCN
Concentration, M	_____	_____	_____

2. Millimoles of ions in the three solutions initially.

	Soln. *c*	Soln. *d*	Soln. *e*
Ag^+	_____	_____	_____
$C_2H_3O_2^-$	_____	_____	_____

3. Volume of titrant used.

	Soln. *c*	Soln. *d*	Soln. *e*
Trial 1	_____	_____	_____
Trial 2	_____	_____	_____

4. Millimoles of Ag^+ ion in solution and in precipitate.

		Soln. *c*	Soln. *d*	Soln. *e*
Solution:	Trial 1	_____	_____	_____
	Trial 2	_____	_____	_____
Precipitate:	Trial 1	_____	_____	_____
	Trial 2	_____	_____	_____

5. Millimoles of acetate ion, $C_2H_3O_2^-$, in solution.

	Soln. *c*	Soln. *d*	Soln. *e*
Trial 1	_____	_____	_____
Trial 2	_____	_____	_____

Report   Name _____ Section _____

6.   Concentration of ions and solubility products.

		Soln. *c*	Soln. *d*	Soln. *e*
$[Ag^+]$:	Trial 1	_____	_____	_____
	Trial 2	_____	_____	_____
$[C_2H_3O_2^-]$:	Trial 1	_____	_____	_____
	Trial 2	_____	_____	_____
K_{sp}:	Trial 1	_____	_____	_____
	Trial 2	_____	_____	_____
	Average K_{sp}	_____	_____	_____

7.   Temperatures of the three solutions.

	Soln. *c*	Soln. *d*	Soln. *e*
Celsius temperature, °C	_____	_____	_____
Absolute temperature, K	_____	_____	_____

8.   Use $\Delta G° = -RT \ln K_{sp}$ to determine the value of $\Delta G°$ at each temperature. ($R = 8.314$ J/mol-deg.)

	Soln. *c*	Soln. *d*	Soln. *e*
$\Delta G°$, kJ/mol	_____	_____	_____

9.   Use $\Delta G° = \Delta H° - T\Delta S°$ a plot of $\Delta G°$s. *T*, and assume both $\Delta H°$ and $\Delta S°$ do not vary with temperature to calculate $\Delta H°$ and $\Delta S°$ by appropriately combining the values of $\Delta G°$ obtained in step 8.

$\Delta H°$, kJ/mol =   _____

$\Delta S°$, J/mol-deg =   _____

Sample Calculations: Part II

Prelab

Name _____ Section _____

Determination of a Solubility Product Constant

1. In Procedure A, step 5, some of the solution containing Ag^+ and $C_2H_3O_2^-$ ions remain with the precipitant in the filter paper. Why doesn't this introduce and error when determining the number of millimoles of Ag^+ in the solution?

2. What is the purpose of the 2 mL of $(NH_4)Fe(SO_4)_2$ that is added to the solution to be titrated?

3. At 25°C, 50.0 mL of 0.150 M $AgNO_3$ is mixed with 30.0 mL of 0.300 M $NaC_2H_3O_2$.

    a. Calculate the initial number of millimoles of Ag^+ and $C_2H_3O_2^-$ present in the solution.

    b. After the solution is filtered to remove the precipitant, a 20.0 mL sample of solution required 12.63 mL of 0.100 M KSCN to titrate to the end point. Determine the number of millimoles of Ag^+ in the 20.0 mL sample.

    c. How many millimoles of Ag^+ must have been in the 80.0 mL of solution?

    d. Determine the number of millimoles of Ag^+ in the precipitant. How many millmoles of $C_2H_3O_2^-$ must be in the precipitant?

Mn (VII) is reduced

$$5e^- + MnO_4^- + 8H^+ \longrightarrow Mn^{2+} + 4H_2O$$ Reduction

(with $+7$ over MnO_4^-, $+2$ over Mn^{2+})

$$\overset{+3}{C_2}O_4^{2-} \xrightarrow{(-8)} 2CO_2 + 2e^-$$ Oxidation

$+6 (-8)$

(Mn (VII) Oxidizing Agent

(also Cr (VI))

$$\overset{+1}{Na_2}\overset{+3}{C_2}\overset{-2}{O_4}$$

$+2 \quad +6 \quad -8$

$$\overset{+3}{C_2}O_4^{2-} \longrightarrow \overset{+4}{2CO_2} + 2e^-$$ Oxidation

$+4 \quad -4$ $C_2O_4^{2-}$ reducing agent

23.3

$Na_2C_2O_4 \qquad KMnO_4$

Part A want M KMnO₄

$$(g\ Na_2C_2O_4)\left(\frac{1mol}{134g\ Na_2C_2O_4}\right)\left(\frac{2mol\ KMnO_4}{5mol\ Na_2C_2O_4}\right) = moles\ KMnO_4$$

★ $$\frac{moles\ KMnO_4}{Vol(L)} = M\ KMnO_4$$

Part B. want g Na₂C₂O₄ (% Na₂C₂O₄)

go backwards + flip

M (KMnO₄) × vol (L) = mols KMnO₄

$$(moles\ KMnO_4)\left(\frac{5mol\ Na_2C_2O_4}{2mol\ KMnO_4}\right)\left(\frac{134g\ Na_2C_2O_4}{1mol}\right) = g\ Na_2C_2O_4$$

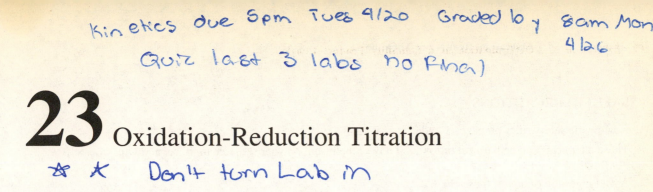

23 Oxidation-Reduction Titration

☆ ★   Don't turn Lab in

INTRODUCTION

In addition to its application to acid-base reactions, titration techniques may be applied to other types of reactions, such as oxidation-reduction. In this experiment, the oxidation of sodium oxalate, $Na_2C_2O_4$, by potassium permanganate, $KMnO_4$, will be studied. Pure $Na_2C_2O_4$ is a solid salt and a very good primary standard. It will be used first to standardize a $KMnO_4$ titrant, which then will be used to determine the percentage purity of an unknown oxalate sample. The reaction is carried out in the presence of dilute sulfuric acid, which prevents the formation of the brown, insoluble MnO_2 and ensures that the reduction of permanganate proceeds according to the following equation:

$$MnO_4^- + 8\ H^+ + 5\ e^- \longrightarrow Mn^{2+} + 4\ H_2O \qquad [23.1]$$

At the same time oxalate ion, $C_2O_4^{2-}$, is oxidized according to the following equation:

$$C_2O_4^{2-} \longrightarrow 2\ CO_2 + 2\ e^- \qquad [23.2]$$

The combination of half-reactions [23.1] and [23.2] yields the overall reaction, equation [23.3].

$$5\ C_2O_4^{2-} + 2\ MnO_4^- + 16\ H^+ \longrightarrow 10\ CO_2 + 2\ Mn^{2+} + 8\ H_2O \qquad [23.3]$$

The permanganate-oxalate reaction system contains a built-in indicator. The end point of the titration is ascertained by noting the presence of the first excess of permanganate ions in the solution being titrated. Thus, when the oxalate has all reacted, the first excess of added permanganate solution will no longer be decolorized via reaction [23.3], and the solution will change from colorless to pink. One drop of excess permanganate solution is sufficient to cause this color change.

At room temperature, the rate of the reaction is too slow to give a sharp, accurate end point. Thus the titration is carried out in the presence of a small quantity of manganese(II) sulfate, $MnSO_4$, which catalyzes the reaction. (*Note*: Mn^{2+} is also a product of the overall reaction. Such a reaction, in which a product acts as a catalyst, is said to be *autocatalyzed*.) Alternatively, the solution may be heated to 60 to 70°C to speed up the reaction.

Solid $KMnO_4$ is usually contaminated with manganese dioxide, MnO_2, which catalyzes the reduction of the permanganate in aqueous solution as follows:

$$4\ MnO_4^- + 2\ H_2O \longrightarrow 4\ MnO_2 + 3\ O_2 + 4\ OH^- \qquad [23.4]$$

Therefore, before the $KMnO_4$ solution can be standardized, any MnO_2 contaminant must be removed. This is done by filtration through glass wool. Filter paper must not be used because $KMnO_4$ is readily reduced by the organic matter. By taking these precautions, however, standard $KMnO_4$ solutions may be stored for months.

The prepared $KMnO_4$ used in this experiment will be prefiltered, and the H_2SO_4 will have $MnSO_4$ added to it prior to their use in this experiment. All titrations are to be carried out with a relative precision of within ±1%. When two successive trials agree within 1%, it is not necessary to perform another trial.

SAFETY PRECAUTIONS

Review the safety rules on pages 1 and 2.

These chemicals can be harmful or toxic. If any of them spill on your skin, clean, rinse, and wipe yourself thoroughly.

Avoid ingestion of any of these chemicals.

Wash your hands before leaving the laboratory.

PROCEDURE

A. Standardization of Prepared Permanganate Solution

Weigh about 0.25 g (to ±0.001 g) of $Na_2C_2O_4$ into a 250 mL beaker. Dissolve the $Na_2C_2O_4$ in about 100 mL of distilled H_2O and acidify with 25 mL of 3 M H_2SO_4, to which some manganese(II) sulfate has been added. You may wish to heat this solution to near boiling before titrating. This has no effect on the titration except to make the reaction go faster. Titrate with the prepared permanganate solution, stirring constantly. The end point is the first persistence of the permanganate color. Be careful not to add the permanganate solution too rapidly initially, as the color will not clear if this is done. One should add no more than 1 mL of permanganate solution before stopping and waiting until the solution clears. The color change in this experiment is not rapid like that of phenolphthalein. Consequently, the titration proceeds more slowly. If your solution turns a permanent muddy brown (MnO_2 formation) rather than pink or clear, you should discard the entire trial and start again.

Use the mass of pure $Na_2C_2O_4$, the mole ratio of $MnO_4^-/C_2O_4^{2-}$ and the mL of $KMnO_4$ required to reach the end point to calculate the molarity of $KMnO_4$ titrant. The calculated molarities should be consistent with a relative average deviation of ±1% or less for two consecutive trials before you go on to the next part.

Ignore for now

B. Analysis of an Unknown Oxalate

Weigh 0.75 g (to ±0.001 g) of the unknown solid into a 250 mL beaker and follow the same procedure as in Part A. Use the mL $KMnO_4$, the mole ratio of $MnO_4^-/C_2O_4^{2-}$, and the average molarity of $KMnO_4$ to calculate the mass and percentage of $Na_2C_2O_4$ in the unknown sample. The calculated percent mass should be consistent with a relative average deviation of ±1% or less for two consecutive trials.

DISPOSAL

Titration solutions: *Dispose* in a waste bottle labeled *heavy metals—solutions*.

Unused $KMnO_4$: *Dispose* in a waste bottle labeled *waste $KMnO_4$*. It will be treated further with reducing agent and safely disposed.

Unused solids: *Dispose* $Na_2C_2O_4(s)$ and $Na_2C_2O_4/Na_2CO_3(s)$ in a waste bottle labeled *inorganic waste solids*.

$$.271 \left(\frac{1\, mol}{134\, g} \right)$$

Report   Name _____ Section _____

DATA

A. Standardization of Solution

Only 1

	Trial 1	Trial 2	Trial 3
Mass of $Na_2C_2O_4$, g	.271g	_____	_____
Moles of $Na_2C_2O_4$	2.02	_____	_____
Initial buret reading, mL	9.2	_____	_____
Final buret reading, mL	29.4	_____	_____
Volume of $KMnO_4$ solution, mL	20.2	_____	_____	0.0202L
Molarity of $KMnO_4$ solution	0.0400	_____	_____
Average molarity, M		_____	
Relative average deviation in M		_____	

B. Analysis of Unknown Oxalate

Unknown number ____5____

	Trial 1	Trial 2	Trial 3
Mass of unknown, g	.750	_____	_____
Initial buret reading, mL	29.4	_____	_____
Final buret reading, mL	39.9	_____	_____
Volume of $KMnO_4$ solution, mL	10.5	_____	_____	0.0105L
Moles of $KMnO_4$	0.00042	_____	_____
Moles of $Na_2C_2O_4$ in unknown	0.00105 mols	_____	_____
Mass of $Na_2C_2O_4$ in unknown, g	0.141 g	_____	_____
Percent of $Na_2C_2O_4$ in unknown	18.8 %	_____	_____
Average percent $Na_2C_2O_4$ in unknown		_____	
Relative average deviation		_____	

Sample Calculations

Report   Name _____ Section _____

QUESTIONS

1. On Part A of the procedure, how would the calculated molarity of $KMnO_4$ be affected (too large, too small or no effect) by the following errors. Explain your answers.

   a. The $KMnO_4$ is added too rapidly and too much $KMnO_4$ is added, overshooting the end point.

   b. Some of the solid Na_2CO_4 is spilled on the bench top after weighing but before beginning the titration.

2. In Part B of the procedure, how would the calculated percent oxalate of the unknown be affected (too large, too small, or no effect) by the following analytical errors. Explain your answers.

   a. The unknown solid sodium oxalate was dissolved in 75 mL of water rather than the 100 mL as described in the procedure.

   b. A few drops of $KMnO_4$ titrant missed the reaction flask containing the unknown sodium oxalate.

   c. When weighing out the solid unknown sodium oxalate, the actual mass is 0.758 grams. However, the mass is mistakenly recorded in the lab notebook as 0.785 grams.

Prelab   Name _____ Section _____

Oxidation–Reduction Titration

1. What does it mean to *standardize* the $KMnO_4$ solution?

2. In the titration reaction, which substance is

   a. reduced?   _____

   b. oxidized?   _____

3. What is the mole ratio of $MnO_4^-/C_2O_4^{2-}$ in the chemical reaction in this experiment?

4. What is the purpose of the 3 M H_2SO_4 solution? What is the purpose of the manganese(II) sulfate?

5. Why is it necessary to titrate slowly in this experiment?

24 Electrolysis: Faraday's Law and Determination of Avogadro's Number

INTRODUCTION

During the electrolysis of an ionic solution, chemical reactions involving electrons occur at the surfaces of the electrodes. For example, consider the simple reduction of hydrogen ion to hydrogen gas at the cathode of an electrolysis cell. This reaction is

$$H^+(aq) + e^- \rightarrow \frac{1}{2}H_2(g) \qquad [24.1]$$

Experimental evidence indicates that one electron reacts with one hydrogen ion, or two electrons react with two hydrogen ions, or one mole of electrons react with one mole of hydrogen ions. The ratio of electrons to hydrogen ions is 1:1, and the ratio of electrons to hydrogen atoms formed is 1:1. However, when O_2 is produced from the electrolysis of water at the anode of a cell according to

$$2\,OH^-(aq) \rightarrow \frac{1}{2}O_2(g) + H_2O + 2\,e^- \qquad [24.2]$$

the formation of $\frac{1}{2}O_2$ occurs with the release of two electrons to the electrochemical circuit. That is, the ratio of product atoms to electrons is 1:2 rather than 1:1, since each oxygen atom loses two electrons during the electrolysis of OH^-.

The discussion may be extended to indicate that by measuring the flow of electrons that enter and leave the solution through the electrodes, one has an indirect count of the number of ions involved in reaction and the number of atoms formed. This relationship is known as Faraday's law, which states that the amount of chemical change produced is proportional to the quantity of electric charge that passes through an electrolytic cell.

Some review of important facts is necessary before we may measure electron flow. The charge on the electron, e^-, has been found to be 1.602×10^{-19} coulomb (C) and is negative by convention. Since the charge of the hydrogen ion is neutralized by one electron, it must be of the same magnitude as the charge on the electron (1.602×10^{-19} C) and it must be opposite in sign (+).

By measuring the number of electrons necessary to produce 1.008 g of hydrogen at an electrode, we indirectly measure the number of hydrogen atoms in 1.008 g of hydrogen. We do this by measuring the number of coulombs that pass through the solution. From this the number of electrons can be calculated.

EXAMPLE: Assume that 9785 C of electricity pass through an electrolysis cell. How many electrons are involved in an electrode process and how many H^+ ions are reduced?

Since each electron carries 1.602×10^{-19} C, the answer to this problem is obtained as follows.

$$\text{No. of electrons} = 9785\,C \times \frac{1\,e^-}{1.602 \times 10^{-19}\,C} \times \frac{1\,H^+}{1\,e^-} = 6.108 \times 10^{22}\,H^+ \text{ ions}$$

Hence, if hydrogen ions were being reduced at the cathode, 9785 C of electricity would reduce 6.108×10^{22} hydrogen ions.

To determine the number of coulombs that pass through a solution we must measure the amperage and the time of passage of electricity through the solution. The *ampere* (A), a unit of electrical current, is a measure of the number of coulombs that pass through a conductor in one second.

EXAMPLE: How many hydrogen ions can be reduced in 10.00 min by a current of 1.265 A?

1.265 A is equivalent to 1.265 C s^{-1}

$$\text{Number of hydrogen ions} = 10 \text{ min} \times \frac{60 \text{ s}}{\text{min}} \times \frac{1.265 \text{ C}}{\text{s}} \times \frac{1 \, e^-}{1.602 \times 10^{-19} \text{C}} \times \frac{1 \text{ hydrogen ion}}{1 \, e^-}$$

$$\text{"} \qquad\qquad = 4.738 \times 10^{21} \text{ hydrogen ions}$$

The object of this experiment is to determine Avogadro's number, the number of atoms in one mole of any element. In addition, the value of the faraday will be determined. The *faraday* (F) is the quantity of electricity associated with the passage of Avogadro's number (6.02×10^{23} or 1 mol) of electrons during an electrochemical process.

SAFETY PRECAUTIONS

Review the safety rules on pages 1 and 2.

Burns can result if H_2SO_4 spills on skin and is not rinsed thoroughly with running water. Any H_2SO_4 (even if dilute) spills must be cleaned up, rinsed, and wiped thoroughly.

Make sure to wear goggles at all times!

Keep your hands off the wire leads and contacts, and out of the solution when the power supply is plugged in.

PROCEDURE

Make sure the power supply is *not* plugged in and assemble the apparatus as shown in Figure 24-1. Fill each eudiometer tube with 1 M H_2SO_4 solution. In addition, fill your largest beaker about two-thirds full of acid solution. Then place one finger over the end of the tube, invert, and dip the end of the tube in the beaker of solution. No air should be in the tubes before electrolysis begins. Rinse your hands immediately and thoroughly with running water.

Complete the circuit by plugging in and switching on the power supply. You should use a current of approximately 250 to 500 milliamperes (mA) to enable you to complete the experiment in a reasonable time. You should also attempt to keep the amperage as constant as possible by adjusting the voltage and current dials of the power supply. Record the amperage in your notebook every minute for the first 5 min and every 3 min thereafter until approximately 80 mL of hydrogen gas has formed. Then stop the experiment.

Stir the solution. Record the temperature of the solution at the end of the experiment. Due to warming of the solution during electrolysis, the solution must be allowed to cool for 3 to 5 min. When two solution temperatures, read 1 min apart, are within 1°C of each other, record the last as the temperature of the gas.

Record the volumes of hydrogen and oxygen gas. Record the height of the liquid column in each eudiometer tube by measuring the difference between the level of the solution in the beaker and in the eudiometer tube. These measurements should be as accurate as possible. (Use a meter stick as shown in Experiment 10.) The measurements should be in mm.

Record the barometric pressure for the day, in mm Hg.

The water vapor pressure above the acid solution is given by the equation

$$p = -5.4 + 1.1t$$

at temperatures near room temperature where t is in °C and p in mm Hg. Calculate and record the vapor pressure above the acid solution.

Do a second determination if time permits.

DISPOSAL

1 M H_2SO_4 solution: Consult with your instructor and do one of the following:
1. Dispose the solution in a labeled waste container.
2. Return the solution to the original bottle for reuse.
3. Add phenolphthalein and neutralize with 6 M NaOH (first appearance of permanent pink color). Flush the resulting salt solution down the sink with running water.

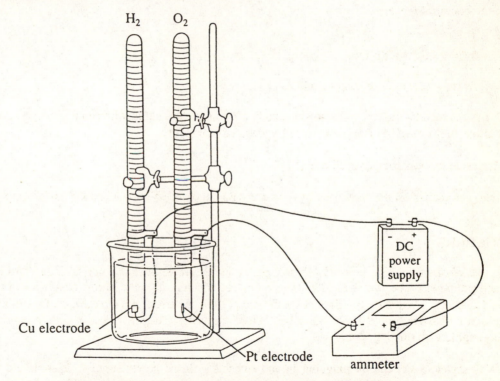

Figure 24-1. Apparatus for the Electrolysis of H_2SO_4

Report   Name _____ Section _____

DATA

TRIAL 1		TRIAL 2	
Time, min	Current, mA	Time, min	Current , mA
_____	_____	_____	_____
_____	_____	_____	_____
_____	_____	_____	_____
_____	_____	_____	_____
_____	_____	_____	_____
_____	_____	_____	_____
_____	_____	_____	_____
_____	_____	_____	_____
_____	_____	_____	_____
_____	_____	_____	_____
_____	_____	_____	_____
_____	_____	_____	_____

	Trial 1	Trial 2
Volume of hydrogen gas, mL	_____	_____
Temperature of solution, °C	_____	_____
Time lapse during passage of current, s	_____	_____
Difference in liquid levels (H_2 tube and beaker), mm	_____	_____
Volume of oxygen gas, mL	_____	_____
Difference in liquid levels (O_2 tube and beaker), mm	_____	_____
Barometric pressure, mm Hg	_____	_____

Report   Name _____ Section _____

                                                      Lab Partner _____

QUESTIONS AND CALCULATIONS

In all calculations, use your data and the introduction to this experiment. Do not use values from textbooks, handbooks, and so forth, unless absolutely necessary.

1. The density of a 1 M H_2SO_4 solution is 1.06 g/mL. Use this density and your recorded data for volumes of hydrogen and oxygen gases, level differences in each tube and vapor pressure of acid solution in each tube to calculate:

   a. the pressures of the *dry* hydrogen and oxygen gases,
   b. the number of moles of each gas,
   c. the mass of each gas.

   Show sample calculations. (Do not use your amperage data for these. Refer to Experiment 10 for calculations. The density of the H_2SO_4 solution must be used in the level-difference correction.)

2. From the calculated number of Coulombs passed through the cell, calculate the number of hydrogen ions reduced during the electrolysis.

3. Use the number of hydrogen ions reduced (Question 2) and the mass of hydrogen (Question 1) to calculate Avogadro's number (the number of hydrogen atoms in 1.008 g of hydrogen).

Report   Name _____ Section _____
                                                        Lab Partner _____

4. Calculate the value of the faraday, the number of coulombs necessary to reduce Avogadro's number of hydrogen ions or produce 1.008 g of hydrogen gas. Use your value of Avogadro's number and the quantity of electricity used.

5. Hydrogen is more nearly an ideal gas than oxygen. What effect does this difference have on the measured volumes of these two gases?

6. Compared with the accepted values in your textbook, calculate the percent error in your values of (a) Avogadro's number, and (b) the Faraday.

7. Give three important sources of error in this experiment. Which of these most significantly affects your results? Explain.

Prelab   Name _____ Section _____

                                                           Lab Partner _____

Electrolysis: Faraday's Law and Determination of Avogadro's Number

1. When inverting the filled eudiometer tubes, why is it important that no air be allowed to enter?

2. Why must you measure the difference in liquid levels in the eudiometer tube? (*Hint*: Refer to Experiment 10.)

3. Suppose a current of 450 mA (milliamperes) is passed through a 1 M H_2SO_4 solution for 40.0 minutes.

    a. How many coulombs of electricity have been passed?

    b. How many H^+ ions have been reduced at the cathode?

4. If 80.0 mL of hydrogen gas is collected at an atmospheric pressure of 770 mm Hg, the water temperature is 21°C, and the liquid level inside the gas collecting tube is 85 mm above that in the beaker, calculate the following:

    a. The pressure of "dry" hydrogen gas in the tube

    b. The number of moles of hydrogen gas in the tube

25

An Investigation of Voltaic Cells - The Nernst Equation

INTRODUCTION

This experiment investigates the electrical potential (voltage) produced by an oxidation-reduction reaction and the effect of concentration on that potential. **Oxidation** is the process in which the oxidation state of an element increases; the element also loses electrons. In **reduction**, the oxidation state of an element decreases; the element gains electrons. In oxidation-reduction reactions, an oxidation must always accompany a reduction; neither process can occur alone.

This experiment has five objectives:

1. Measure the potentials of cells formed with a variety of different solutions and electrodes.
2. Calculate half reaction potentials using the Cu^{2+}/Cu half reaction as a reference.
3. Measure the effect of changes in concentration on cell potential.
4. Use the Nernst Equation to relate concentration and cell potential.
5. Calculate the equilibrium constant for a dissociation reaction using the equilibrium concentration of Cu^{2+} determined from a measurement of potential.

When an iron nail is placed in a copper (II) nitrate solution and left to stand, the blue color of the solution fades and a brown-red deposit forms on the nail. If the brown-red deposit is scraped away, the nail is found to be somewhat smaller and to have an uneven surface. Chemical analysis reveals that the brown-red deposit is copper metal and Fe^{2+} ions are in solution. Some elemental iron, Fe, has become Fe^{2+} ions, and some Cu^{2+} ions have become elemental copper, Cu. These changes are summarized by the following two half-equations.

$$Fe(s) \longrightarrow Fe^{2+}(aq) + 2\ e^- \tag{25.1}$$

$$Cu^{2+}(aq) + 2\ e^- \longrightarrow Cu(s) \tag{25.2}$$

Equation [25.1] represents an oxidation half-reaction, and equation [25.2] represents a reduction half-reaction. Another way of stating this is that Fe has been oxidized to Fe^{2+} and Cu^{2+} has been reduced to Cu.

Yet another way to summarize the results of the observation described above is to write an overall equation. Two different types of equations may be written:

$$Fe(s) + Cu(NO_3)_2(aq) \longrightarrow Fe(NO_3)_2(aq) + Cu(s) \tag{25.3}$$

$$Fe(s) + Cu^{2+}(aq) \longrightarrow Fe^{2+}(aq) + Cu(s) \tag{25.4}$$

Equation [25.3] represents the complete reaction for these observations. If the colorless solution produced by the reaction were gently evaporated, solid $Fe(NO_3)_2$ and Cu would be obtained; little or no $Cu(NO_3)_2$ would be found in the now colorless solution. Equation [25.4] summarizes the essential features of the reaction and eliminates the ion that is unchanged in the process (NO_3^- in this case). Equation [25.4] is called the **net ionic equation** for this reaction.

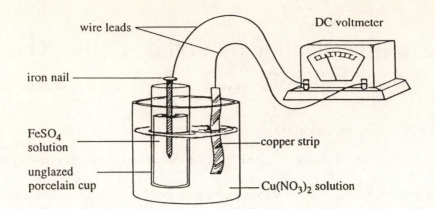

Figure 25-1. Apparatus for Measurement of Cell Potentials

Half-reaction equations can be used to determine the balanced oxidation-reduction reaction. If electron gain and loss are balanced, the two half-equations can be added to yield the net ionic equation for the overall reaction.

When studying oxidation-reduction reactions it is convenient to separate the two half-reactions from one another. This is accomplished using the experimental setup shown in Figure 25-1. The porous cup is necessary to complete the electric circuit by allowing electrical contact between the two solutions while keeping the solutions separate from one another. If the $Cu(NO_3)_2$ solution were allowed to come into contact with the iron nail, reaction [25.4] would occur directly. If they are separated by the porous cup, the electrons produced by half-reaction [25.1] travel from the iron nail through the wire to the copper strip where half-reaction [25.2] occurs. Reaction [25.4] is the resulting overall reaction.

The setup of Figure 25-1 can be used to measure the voltage, or potential, of the cell reaction [25.4]. This potential changes if a zinc strip is used in place of the iron nail and $ZnSO_4$ solution is used in place of the $FeSO_4$ solution. Each half-reaction has a distinct potential associated with it and the potential of a cell is the sum of these half-cell potentials.

The Nernst Equation

The concentration of the ions in solution also affects the potential of the half-cell. For the reaction

$$Fe^{3+}(aq) + e^- \longrightarrow Fe^{2+}(aq) \qquad [25.5]$$

the half-cell potential is given by the Nernst equation

$$E_{red} = E°_{red} - \frac{0.0592}{n} \log \frac{[Fe^{2+}]}{[Fe^{3+}]} \qquad [25.6]$$

where $E°_{red}$ is the standard reduction half-cell potential (the voltage measured when $[Fe^{2+}] = [Fe^{3+}] = 1.00$ M) and n is the number of electrons transferred in the half-reaction. In the case of reaction [25.5], n = 1. For half-reaction [25.2], the Nernst equation is

$$E_{red} = E°_{red} - \frac{0.0592}{2} \log \frac{1}{[Cu^{2+}]} \qquad [25.7]$$

Frequently, it is easier to use the Nernst equation for the entire cell. For reaction [25.4], the Nernst equation is (where $E°_{cell} = E°_{ox} + E°_{red}$)

$$E_{cell} = E°_{cell} - \frac{0.0592}{2} \log \frac{[Fe^{2+}]}{[Cu^{2+}]}$$ [25.8]

Concentration Cells

The Nernst equation also may be used to determine the potential of a cell in which the net reaction is one of simple dilution.

$$Cu^{2+} (1.0 \text{ M}) \longrightarrow Cu^{2+} (0.050 \text{ M})$$ [25.9]

For this reaction $E°_{cell} = 0.00$ V (since $E°_{Ox} = E°_{Red}$) and the cell voltage depends only on concentration.

$$E_{cell} = 0.00 - \frac{0.0592}{2} \log \frac{0.050}{1.0} = 0.0385 \text{ V}$$

In general, the Nernst equation contains the molar concentrations of the concentrated and dilute solutions, which are physically separated in the cell by a porous cup or a salt bridge.

$$E_{cell} = -\frac{0.0592}{n} \log \frac{[M^{n+}(dil)]}{[M^{n+}(conc)]}$$ [25.10]

Equation [25.10] is the Nernst equation for the reaction and cell shown below.

$$M^{n+}(conc) \longrightarrow M^{n+}(dil) \qquad M \mid M^{n+}(dil) \parallel M^{n+}(conc) \mid M$$ [25.11]

SAFETY PRECAUTIONS

Review the safety rules on pages 1 and 2.

Be very careful in handling chemicals. Wipe all spills promptly with a damp sponge or towel. If any chemicals get on your hands or clothing, wash immediately with soap and large volumes of running water. If more than a few drops of chemical solution spill on you, tell your instructor while you are washing off the spill. Wash your hands carefully and thoroughly at the end of the experiment before leaving the laboratory.

Use the bromine solution under the hood. Bromine fumes are an irritant. If bromine solution is spilled, call your laboratory instructor.

PROCEDURE

A. Oxidation-Reduction Reactions

1. Add 1.0 M $Cu(NO_3)_2$ solution to two small test tubes to a depth of 1 cm. Partially immerse a clean iron nail in one test tube and a strip of zinc in the other. Observe the color of the solutions and the metals immediately after the metals are immersed.

2. Allow the two test tubes to stand while you set up the apparatus shown in Figure 25-1 for the measurement of cell potentials. Note the color of the solutions every five minutes during the following 15-min period.

3. Remove the metals and gently scrape and wash off the metallic deposit. Describe the color and the texture of the metallic deposit as well as the surfaces of the iron nail and the zinc strip. Especially note any differences between the part of the metal that was immersed in the $Cu(NO_3)_2$ solution and the part that was not.

B. Cell Voltages

1. Place approximately 30 mL of 1.0 M $Cu(NO_3)_2$ solution in a 50 mL beaker.

2. Place the porous cup in the beaker and fill the cup with 1.0 M $FeSO_4$ solution to about the same level as the solution in the beaker, about 10 mL of solution.

3. Place the iron nail in the $FeSO_4$ solution and a copper strip in the $Cu(NO_3)_2$ solution. Connect the red lead from the positive terminal of the voltmeter to the copper electrode using the alligator clip at the end of the lead. Connect the black lead to the iron nail electrode. Do not allow the alligator clips to come into contact with the solutions. You will use the same copper electrode and $Cu(NO_3)_2$ solution throughout the experiment. Only the electrode and solution in the porous cup will change.

4. Allow the electrodes to sit in solution until the voltmeter reading is stable to ±0.01V (fluctuation in the third digit after the decimal will not be significant in this experiment). Read the voltage to the nearest 0.01 V. Repeat the measurement two times by removing the electrodes from solution for a few seconds and then returning them to the solutions. Record the average voltage

5. A voltaic cell undergoes a spontaneous reaction and will therefore always have a positive potential. The alligator clips should be connected to the electrodes such that a positive potential reading is obtained on the voltmeter. The electrode connected to the positive terminal of the voltmeter is positive and is therefore the cathode. In this first case the copper electrode is the positive cathode. The other electrode is negative and is therefore the anode. For each cell record the polarity (+ or −) of the copper electrode.

6. Disconnect the alligator clip from the iron nail. Remove the iron nail and the porous cup from the beaker, discard the solution in the porous cup as described in the disposal section. Thoroughly wash the porous cup and rinse it with distilled water.

7. Repeat steps 2 through 6 with each of the solution-electrode pairs (given in Table 25.1) placed in turn in the porous cup. Each electrode should be long enough (at least 8 cm) so that it is not completely immersed in the solution. If the electrode is not long enough, it can either be hooked over the edge of the porous cup or the alligator clip can be used to hold the top of the electrode out of the solution. The porous cup should be thoroughly rinsed with water followed by rinsing with about 2 mL of solution before the cup is filled with the solution.

Table 25-1. Electrodes to Be Investigated

Solution	Electrode
1. 1.0 M $FeSO_4$	iron nail
2. 1.0 M $ZnSO_4$	Zn metal
3. 1.0 M $MgSO_4$	Mg metal
4. 0.50 M $Pb(NO_3)_2$	Pb metal
5. 0.50 M $Sn(NO_3)_2$	Sn metal
6. 1.0 M $FeSO_4$ and 1.0 M $Fe_2(SO_4)_3$	graphite*
7. 0.1 M NaBr and $Br_2(l)$	graphite*
8. 0.1 M $AgNO_3$	Ag wire

*The graphite electrode serves only for electrical contact with the solution. It does not participate in the oxidation-reduction reaction.

C. Effect of Concentration

Note: In steps 1, 2, and 3, place 1.0 M $Cu(NO_3)_2$ solution in the 50 mL beaker with a copper strip electrode.

1. Prepare 50 mL of 0.10 M $Cu(NO_3)_2$ solution by diluting 5.0 mL of 1.0 M $Cu(NO_3)_2$ solution to 50 mL with distilled water in a 100 mL graduated cylinder. Mix well. Use this solution and a copper strip in the porous cup. Record the potential of the cell and polarity (+ or −) of each electrode. Save as much of this solution as possible for step 4.

2. In like fashion, prepare 50 mL of 0.010 M $Cu(NO_3)_2$ solution from 5.0 mL of 0.10 M $Cu(NO_3)_2$ solution and use it in the porous cup with a copper electrode. Record the potential of the cell and the polarity of the electrodes.

3. Similarly, prepare 50 mL of 0.0010 M $Cu(NO_3)_2$ solution from 5.0 mL of 0.010 M $Cu(NO_3)_2$ solution and use it in the porous cup with a copper electrode. Record the potential of the cell and the polarity of the electrodes.

4. Prepare an additional 50 mL of 0.10 M $Cu(NO_3)_2$ solution using the method described in step 1. Place 0.10 M $Cu(NO_3)_2$ solution in both the beaker (30 mL of solution) and the porous cup (10 mL of solution) and use copper electrodes. Record the cell potential. Add 2.0 mL of 6 M NH_3(aq) to the porous cup. Record the cell potential, the polarity of the electrodes, and the total volume of the solution in the porous cup. The NH_3 reacts with the Cu^{2+} ion, which is blue in solution, to form the $[Cu(NH_3)_4]^{2+}$ complex ion, which is blue-violet in solution.

DISPOSAL

NaBr-Br$_2$ solution: *Dispose in a waste bottle labeled waste bromine/bromide.*
Other Solutions: *Dispose in a waste bottle labeled heavy metals—solutions.*

CALCULATIONS

A. Half Cell Potentials: Use the known standard potential (0.34 V) of the copper half reaction to calculate the potentials of the other indicated half reactions. If the copper electrode was positive, then it behaved as the cathode and underwent reduction.

$$
\begin{array}{lll}
Cu^{2+} + 2e^- \longrightarrow Cu & +0.34\ V \\
\underline{M \qquad\qquad \longrightarrow M^{2+} + 2e^-} & \underline{\quad ? \quad} \\
Cu^{2+} + M \longrightarrow Cu + M^{2+} & \text{Measured potential}
\end{array}
$$

If the copper electrode was negative, then copper was the anode and oxidation occurred at the copper electrode.

$$
\begin{array}{lll}
M^{2+} + 2e^- \longrightarrow M & ? \\
\underline{Cu \qquad\qquad \longrightarrow Cu^{2+} + 2e^-} & \underline{-0.34\ V} \\
M^{2+} + Cu \longrightarrow Cu^{2+} + M & \text{measured potential}
\end{array}
$$

The potential requested in Part A of the questions (p. 225) is for the reduction half reaction. If you calculate the potential for an oxidation half-reaction using the above method, the sign must be changed when recording the value in the table. The last two entries in Part A ask for the potential when the ion concentration is 1 M. This is the $E°_{cell}$ and may be calculated using the Nernst equation since the potential (E_{cell} at 0.1 M) and ion concentration are known for the reaction under the conditions used in the experiment.

B. In Section A of Part B of the questions (p. 225) the Cu^{2+} concentration in the porous cup is determined from the measured potential using the Nernst equation. The calculated value is then compared to the assumed concentration based on the dilutions made when preparing the solutions.
In Section b of Part B of the questions (p. 226) the equilibrium constant for the formation of $Cu(NH_3)_4^{2+}$ (shown below) is calculated based on the initial concentration of each species and equilibrium concentration of Cu^{2+} (see Experiment 16 for a review of calculating equilibrium constants). The equilibrium concentration of Cu^{2+} is determined from the measured potential using the Nernst equation.

$$Cu^{2+} + 4\,NH_3 \leftrightarrows Cu(NH_3)_4^{2+}$$

C. Calculate the $E°_{cell}$ for the reactions shown. Use the half-cell potentials from Part A of the Questions.

Prelab   Name _____ Section _____

An Investigation of Voltaic Cells–the Nernst Equation

1. Why is it important not to allow the alligator clips to dip into the solutions in Part B, step 3? What would happen if an alligator clip did touch the solution?

2. A voltaic cell is constructed in which a copper wire is placed in a 1.0 M $Cu(NO_3)_2$ solution and a strip of gold is placed in a 1.0 M $AuNO_3$ solution. The measured potential of the cell is found to be 1.36 V and the copper electrode is negative. The $E°_{Red}$ for the Cu^{2+}/Cu half-cell is +0.34 V.

    a. Since the copper electrode is negative, oxidation takes place at the copper electrode. Write the oxidation half-reaction for the Cu^{2+}/Cu half-cell and indicate its standard oxidation potential, $E°_{Ox}$.

    b. Write the half-reaction for the Au^+/Au half-cell.

    c. Write the overall balanced oxidation-reduction and indicate the standard cell potential $E°_{cell}$.

    d. Calculate the standard reduction potential, $E°_{Red}$, for the Au^+/Au half-cell.

Report   Name _____ Section _____

DATA

A. Oxidation-Reduction Reactions

1. Describe the appearance of the following metals and solutions immediately after the metals are immersed.

   a. Zn test tube

   b. Fe test tube

2. Describe the appearance of the metals and solutions and the deposit with respect to time.

Time	5 min	10 min	15 min

   a. Zn solution

   b. Fe solution

   c. Zn metal

   d. Fe metal

Report   Name _____ Section _____

B. Cell Voltages

For each cell you constructed, record (a) the potential (voltage) of the cell, (b) the polarity of the copper electrode, and (c) the net ionic equation for the reaction that occurred.

Cell	Average Potential (V)	Copper Polarity (+ or −)	Net Ionic Equation
1	_____	_____	_____
2	_____	_____	_____
3	_____	_____	_____
4	_____	_____	_____
5	_____	_____	_____
6	_____	_____	_____
7	_____	_____	_____
8	_____	_____	_____

C. Effect of Concentration

For each part, record the potential of the cell and the polarity of the electrode in the porous cup.

Part	Contents of Porous Cup	Cell Potential (V)	Porous Cup Electrode Polarity , (+ or −)
1	0.10 M $Cu(NO_3)_2$(aq)	_____	_____
2	0.010 M $Cu(NO_3)_2$(aq)	_____	_____
3	0.0010 M $Cu(NO_3)_2$(aq)	_____	_____
4	0.10 M $Cu(NO_3)_2$(aq) + 6 M NH_3(aq)	_____	_____

Report

Name _____ Section _____

QUESTIONS

A. Half-Cell Potentials

The standard potential ($E°$) of the half-reaction $Cu^{2+} + 2\,e^- \rightarrow Cu$ is 0.34 V. Determine the potential of each of the following half-reactions at the indicated concentrations. Use only your data (no data from tables please). (*Hint*: Some half-reactions may have negative potentials. You may need to use the Nernst equation.)

Half-Reaction	Concentration	Reduction Potential
$Fe^{2+} + 2\,e^- \rightarrow Fe$	1.0 M	
$Zn^{2+} + 2\,e^- \rightarrow Zn$	1.0 M	
$Mg^{2+} + 2\,e^- \rightarrow Mg$	1.0 M	
$Pb^{2+} + 2\,e^- \rightarrow Pb$	0.5 M	
$Sn^{2+} + 2\,e^- \rightarrow Sn$	0.5 M	
$Fe^{3+} + e^- \rightarrow Fe^{2+}$	1.0 M	
$\frac{1}{2}\,Br_2(l) + e^- \rightarrow Br^{-\,*}$	0.1 M	
$Ag^+ + e^- \rightarrow Ag$	0.1 M	
$Ag^+ + e^- \rightarrow Ag$	1.0 M	
$\frac{1}{2}\,Br_2(l) + e^- \rightarrow Br^-$	1.0 M	

*Note: $Br_2(l)$ is considered to be a pure phase and its concentration does not appear in the Nernst equation.

B. Calculation of Concentration

a. Use the cell potentials of Part C (Data) and the Nernst equation to determine the Cu^{2+} concentration in the porous cup. Then compare it with the assumed concentration, and calculate the percent error in the assumed concentration.

Part	Assumed [Cu^{2+}]	Experimental [Cu^{2+}]	% Error
1	0.10 M		
2	0.010 M		
3	0.0010 M		
4	unknown		

Report   Name _____ Section _____

b. Using the data from the procedure Part C.4 (page 221) calculate the $[Cu^{2+}]$ and $[Cu(NH_3)_4]^{2+}$ in the porous cup and determine the value of the formation constant of the $[Cu(NH_3)_4]^{2+}$ complex, K_f, where

$$K_f = \frac{[(Cu(NH_3)_4)^{2+}]}{[Cu^{2+}][NH_3]^4}$$

C. Cell Potentials

Calculate the potential of each of the following cells, using your half-cell potentials from Part A of Results. Note that the equations are not balanced.

$Fe^{2+} + Zn \rightarrow Zn^{2+} + Fe$ _____

$Pb^{2+} + Mg \rightarrow Mg^{2+} + Pb$ _____

$Sn + Fe^{3+} \rightarrow Fe^{2+} + Sn^{2+}$ _____

$Ag^+ + Pb \rightarrow Ag + Pb^{2+}$ _____

$Br_2 + Zn \rightarrow Br^- + Zn^{2+}$ _____

26 Determination of Water Hardness

INTRODUCTION

Water that contains certain dissolved minerals in relatively large amounts is said to be *hard*. Although hard water is not necessarily unhealthy, there are two common problems associated with it. First, hard water contains certain cations, such as Ca^{2+}, which form water-insoluble compounds with soaps. The formation of these precipitates reduces the cleaning ability of the soap and, at the same time, results in the formation of unsightly "scum" on clothing and in tubs and sinks. If Fe^{3+} is present, a red-brown deposit might form on the surfaces of sinks and tubs. Second, hard water encourages the buildup of boiler scale in water heaters, pipes, etc., which can cause considerable damage to expensive equipment and require time and expense for periodic cleaning.

The major ions responsible for natural water hardness are Ca^{2+}, Mg^{2+}, Fe^{3+}, and HCO_3^-. The presence of bicarbonate ion is classified as *temporary* hardness because the water may be "softened" by boiling. Heating removes some of the HCO_3^- as $CO_2(g)$, but also produces boiler scale when CO_3^{2-} ion is formed and precipitates with Ca^{2+}, Mg^{2+}, and Fe^{3+} ions. The equilibria that occur in an aqueous solution of carbonate ion are summarized in the following chemical equations, and the net equation [26.1] represents the heating reaction that can serve to reduce temporary hardness.

$$HCO_3^-(aq) + H_2O \rightleftharpoons H_3O^+ + CO_3^{2-}(aq)$$

$$HCO_3^-(aq) + H_3O^+ \rightleftharpoons H_2O + H_2CO_3(aq)$$

$$H_2CO_3(aq) \rightleftharpoons H_2O + CO_2(g)$$

Net equation: $2\,HCO_3^-(aq) \rightleftharpoons H_2O + CO_3^{2-}(aq) + CO_2(g)$ [26.1]

Water that contains ions that are not removed by heating, such as Ca^{2+}, Mg^{2+}, Fe^{3+}, and SO_4^{2-}, is said to be *permanently hard*.

The purpose of this experiment is to determine the concentration of permanent hardness in water by titration with a standard solution of ethylenediaminetetraacetic acid (EDTA). This acid acts as a polydentate ligand for most metal cations in aqueous solutions. Most of the complexes formed are quite stable and, therefore, the use of EDTA as a titrant in chemical analysis has found widespread application. The structural formula of the tetraprotic EDTA molecule follows:

$$\text{HOOC} - \text{CH}_2 \qquad\qquad \text{CH}_2 - \text{COOH}$$
$$\ddot{N} - \text{CH}_2 - \text{CH}_2 - \ddot{N}$$
$$\text{HOOC} - \text{CH}_2 \qquad\qquad \text{CH}_2 - \text{COOH}$$

Often the term H_4Y is used to indicate the above substance. Because H_4Y is a polyprotic acid, several ionization equilibria are involved. The most important are as follows:

$$H_4Y \rightleftharpoons H^+ + H_3Y^- \qquad K_{a1} = 1.00 \times 10^{-2} \qquad\qquad [26.2]$$

$$H_3Y^- \rightleftharpoons H^+ + H_2Y^{2-} \qquad K_{a2} = 2.16 \times 10^{-3} \qquad\qquad [26.3]$$

$$H_2Y^{2-} \leftrightarrows H^+ + HY^{3-} \qquad K_{a3} = 6.92 \times 10^{-7} \qquad\qquad [26.4]$$

$$HY^{3-} \leftrightarrows H^+ + Y^{4-} \qquad K_{a4} = 5.50 \times 10^{-11} \qquad\qquad [26.5]$$

Because the parent acid H_4Y is only sparingly soluble in water, the relatively soluble and readily available disodium salt ($Na_2H_2Y \cdot 2\,H_2O$) serves as the starting material for the preparation of standard EDTA solutions used in titrimetry. Since the predominant species in a solution of this salt is the H_2Y^{2-} ion, the pH of the resulting solution of this salt is approximately $\frac{1}{2}(pK_2 + pK_3)$, or 4.40.

The distribution of EDTA among its undissociated and dissociated forms varies considerably with pH. At any particular pH, the distribution of EDTA species may be calculated from the four acid dissociation constants. These calculations can be plotted graphically as a distribution diagram. For example, by using the EDTA distribution diagram in Figure 26-1, which plots the fractional amount of each species as a function of pH, it can be seen that in the 9-12 pH range only Y^{4-} and HY^{3-} are present in significant concentrations, and in the 6-9 pH range, H_2Y^{2-} and HY^{3-} predominate.

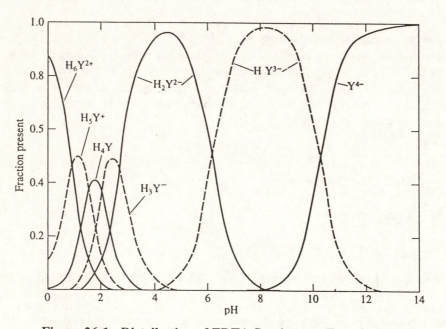

Figure 26-1. Distribution of EDTA Species as a Function of pH

The Y^{4-} ion forms very stable, one–to–one complexes with practically every metal ion, depending on pH and other conditions. Above pH = 12, the Y^{4-} species predominates, and it is available to coordinate metal cations.

$$M^{n+} + Y^{4-} \leftrightarrows MY^{n-4} \qquad\qquad [26.6]$$

An example of a six-coordinated species of an octahedral central metal ion and an EDTA molecule is the cobalt(III)-EDTA complex, CoY^-, whose structure is shown in Figure 26-2.

Most metal ion–EDTA titrations are performed in neutral or alkaline solutions because, although metal-EDTA complexes formed in titrations are quite stable, they can undergo dissociation in the presence of acid, which causes the equilibrium to shift away from formation of the complex.

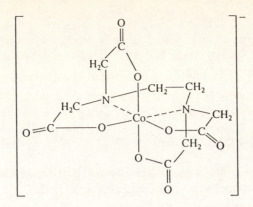

Figure 26-2. Cobalt(III)-EDTA Complex Ion

This equilibrium (equation 26.7) will lie to the left if the pH is maintained at a neutral or alkaline level.

$$MY^{n-4} + 2\,H^+ \rightleftharpoons M^{n+} + H_2Y^{2-} \tag{26.7}$$

As can be seen from Figure 26-1, the predominant species in neutral and slightly alkaline solutions are H_2Y^{2-} and HY^{3-}, and the net titration reactions can be written as in the following two equations:

$$M^{n+} + H_2Y^{2-} \rightleftharpoons MY^{n-4} + 2\,H^+ \tag{26.8}$$

$$M^{n+} + HY^{3-} \rightleftharpoons MY^{n-4} + H^+ \tag{26.9}$$

The precise titration reaction, of course, depends upon the exact pH of the solution. The liberation of H^+ ions during titration would cause the pH to decrease and adversely affect the formation of the metal–EDTA complex. Therefore, in practice, a large excess of an inert buffer system, such as NH_3/NH_4Cl, is used to keep the pH near the desired value for the particular application. In addition to maintaining the desired pH, the NH_3 system forms stable complexes with some metal ions, such as Cd^{2+}, Cu^{2+}, Ni^{2+}, and Zn^{2+}, and thus serves to prevent the undesired precipitation of these ions as metal hydroxides.

A very large number of indicators has been studied and employed for complexometric titrations. The term *metallochromic indicator* is frequently used to describe these substances because they form stable, highly colored complexes with most of the metal ions of interest. They are acid-base indicators as well, and the combination of their metal ion and acid–base indicating properties is used to obtain the indicator behavior desired in complexometric titrations. For example, Calmagite (and Eriochrome Black T) forms a wine-red complex with most metal ions, such as Ca^{2+}.

$$Ca^{2+} + Calmagite \rightleftharpoons Ca(Calmagite)^{2+} \tag{26.10}$$

If EDTA is added to this solution, the Ca^{2+} is gradually removed from the Ca-Calmagite complex and forms the more stable CaY^{2-} complex ion

$$\underset{\textit{wine-red}}{Ca(Calmagite)^{2+}} + Y^{4-} \rightleftharpoons CaY^{2-} + \underset{\textit{blue}}{Calmagite} \tag{26.11}$$

Thus the uncomplexed free form of the Calmagite increases in concentration as it is displaced by the EDTA titrant. When almost no Ca-Calmagite complex remains, the pure blue color of the free Calmagite predominates and denotes the end point. (The end-point color change is sharpened by the presence of Mg^{2+} ion.)

SAFETY PRECAUTIONS

None of the reagents in this procedure is particularly dangerous. However, the normal safety precautions (see pages 1 and 2) should be followed. Wear eye protection and *never* pipet by mouth. Always use a suction device when drawing a solution into a pipet.

PROCEDURE

A. Standardization of 0.01 M EDTA Titrant

1. Rinse a 25.00 mL pipet with small portions of standard 0.01000 M Ca^{2+} solution, and then pipet a 25.00 mL aliquot of the solution into a clean 250 mL Erlenmeyer flask.

2. Add 10 mL of NH_3–NH_4Cl buffer solution from a graduated cylinder and dilute the mixture to approximately 75 mL with distilled water.

3. Add 5 drops of Calmagite indicator solution.

4. Rinse a 50-mL buret with small (5 or 10 mL) portions of the EDTA solution. Then fill it with EDTA solution, and record the initial position of the meniscus. Read the volume to the nearest 0.01 mL.

5. Titrate the Ca^{2+} solution with EDTA until the wine-red color of the Ca indicator complex changes from purple-red to the *pure blue* color of the free form of the indicator. If you are not certain of the color change, go to the window and use natural sunlight to see it or compare it to the color of a titrated blank sample (see note below). Add titrant in drops and fractions of a drop until the color change from wine-red to *pure* blue is permanent. Record the final position of the meniscus after waiting 30 sec for the buret to drain.

   Note: Sometimes it is difficult to tell if the exact end-point color has been reached. The actual color at the end point can be seen if you make a blank sample containing 90 mL of distilled water, 10 mL of buffer, and 5 drops of indicator solution. You can then add several drops of titrant, and if the water is not contaminated with minerals, the color formed will be the blue of the free, uncomplexed indicator.

6. Titrate at least three samples of the standard Ca^{2+} solution with EDTA. Calculate the concentrations of EDTA as follows. In all titrations using EDTA, only one EDTA molecule is used for each metal ion. Thus

$$\text{Molarity of EDTA} = \frac{(\text{Molarity of } Ca^{2+})(25.00 \text{ mL})}{\text{mL EDTA used}}$$

B. Determination of Permanent Hardness of a Water Sample (as ppm $CaCO_3$)

1. Thoroughly rinse the pipet with the water sample and then pipet exactly 50.00 mL of the water sample into a clean 250 mL Erlenmeyer flask. (If the sample is not very hard, a volume of 100.0 mL, measured in a graduated cylinder, may be used.)

2. Add 10 mL of buffer solution (which brings the pH to about 10) and enough indicator to get a distinct color (approximately 10 drops).

3. Refill the 50 mL buret with more of the EDTA titrant used in the standardization titrations and record the initial volume.

4. Titrate the sample solution with EDTA solution until the color changes form purple-red to the first color of pure blue. Be sure to swirl the solution after each addition of titrant. Also, near the end point, rinse down the walls of the flask with distilled water and add the EDTA dropwise.

5. Repeat the titration on a second 50.00 mL sample of water.

DISPOSAL

Solutions: All solutions are non hazardous and may be flushed down the sink with running water.

Hardness Calculation

It is not convenient to determine the concentration of each cation separately, so assume that all the hardness is due to the presence of only Ca^{2+} ions, in the form of calcium carbonate, $CaCO_3$. The unit of hardness may be expressed in terms of molarity of $CaCO_3$, but usually it is expressed as mg $CaCO_3$ per liter of water. This is equivalent to the parts per million of $CaCO_3$. (mg $CaCO_3$/L = parts per million $CaCO_3$ = ppm $CaCO_3$)

The hardness equals:

$$\text{molarity } Ca^{2+} = \frac{\text{mmol } Ca^{2+} \text{ ion}}{\text{mL}} = \frac{(\text{Molarity of EDTA})(\text{mL EDTA added})}{50.00 \text{ mL of water sample}}$$

Since ppm $CaCO_3$ = # mg $CaCO_3$/L,

$$\text{ppm } CaCO_3 = (\text{Molarity } Ca^{2+}) \times \frac{1 \text{ mol } CaCO_3}{1 \text{ mol } Ca^{2+}} \times \frac{100.1 \text{ g } CaCO_3}{1 \text{ mol } CaCO_3} \times \frac{10^3 \text{ mg}}{1 \text{ g}}$$

Prelab   Name _____ Section _____

Determination of Water Hardness

1. What will be the color of the indicator in the titration solution

    a. at the start of the titration?

    b. at the end point?

2. Look again at Figure 26-1. At a pH = 6, which form(s) of EDTA is/are predominantly present in the solution?

3. When temporarily hard water is boiled,
        a. what gas is evolved?

        b. what solid product is formed?

4. 25.00 mL of 0.01100 M Ca^{2+} is titrated to a Calmagite end point with EDTA solution. If the pure blue end-point color occurs at 24.50 mL, what is the molarity of the EDTA?

5. A 50.00-mL water sample requires 20.75 mL of 0.01050 M EDTA to reach a Calmagite end point. Calculate the hardness of this sample in units of ppm $CaCO_3$.

Report   Name _____ Section _____

DATA

A. Standardization of EDTA Solution

Molarity of Ca^{2+} solution = _____ M

	Trial 1	Trial 2	Trial 3
Initial buret reading (±0.01 mL)	_____	_____	_____
Final buret reading (±0.01 mL)	_____	_____	_____
mL of EDTA used	_____	_____	_____
Molarity of EDTA	_____	_____	_____
Average molarity of EDTA		_____	
Relative average deviation		_____	

B. Hardness Titration

	Trial 1	Trial 2
mL of water sample	_____	_____
Initial buret reading (±0.01 mL)	_____	_____
Final buret reading (±0.01 mL)	_____	_____
mL of EDTA used	_____	_____
ppm $CaCO_3$ (mg/L)	_____	_____
Average ppm $CaCO_3$	_____	

Sample Calculations

Report   Name _____ Section _____

QUESTIONS

1. If the pH became too low during the titration of an unknown by EDTA, how would the titration reaction's equilibrium be affected?

2. Water samples A and B are identical except that B contains less iron(III) ion. How would the value of ppm $CaCO_3$ of A compare to that of B (larger, smaller, the same)? Explain.

3. In step 4 of the procedure to standardize the EDTA solution, a student forgets to rinse the buret with a small portion of EDTA solution. (The buret contains water droplets from washing.) If this incorrect concentration of EDTA is used to calculate the ppm $CaCO_3$ of a hard water sample, will the resulting value be too high, too low, or correct? Explain briefly.

4. Why does the Calmagite indicator change from red to blue when EDTA is added to a solution of cations? Describe what is responsible for the color change.

27
The Solvay Process: Preparation of NaHCO$_3$ and Na$_2$CO$_3$

INTRODUCTION

Two of the most abundant commercially produced chemicals that have many familiar and practical uses are NaHCO$_3$ and Na$_2$CO$_3$. Sodium bicarbonate, NaHCO$_3$, is commonly known as baking soda and has many household uses. Sodium carbonate, Na$_2$CO$_3$, commonly known as soda ash, is used as a water softener and in the manufacture of glass, paper, soap, and other chemicals. Na$_2$CO$_3$ is the most important manufactured compound of sodium.

Both compounds may be regarded as salts of carbonic acid, H$_2$CO$_3$. NaHCO$_3$ is the acid salt in which one hydrogen atom has been replaced by a sodium atom; hence it is a salt but retains acidic characteristics. Na$_2$CO$_3$ is the normal salt of H$_2$CO$_3$ in which both hydrogens have been replaced by sodiums. Due to hydrolysis of the CO$_3$$^{2-}ion, Na_2CO_3$ is a basic substance in water. It is produced from NaHCO$_3$ by thermal decomposition.

$$2 \text{ NaHCO}_3 \xrightarrow{300°C} \text{Na}_2\text{CO}_3 + \text{H}_2\text{O} + \text{CO}_2 \qquad \text{[27.1]}$$

Sodium bicarbonate is produced commercially by the Solvay process, utilizing many principles essential to industrial chemical production.

1. Cost effectiveness, involving inexpensive raw materials (limestone, that is, CaCO$_3$, and NaCl).

2. Recycling of reusable by-products (CO$_2$ and NH$_3$).

3. Conversion of other by-products into more usable ones (CaCl$_2$).

4. Use of a simple chemical principle, in which NaHCO$_3$ is separated by fractional crystallization.

The limestone is heated to generate carbon dioxide.

$$\text{CaCO}_3 \xrightarrow{900°C} \text{CaO} + \text{CO}_2 \qquad \text{[27.2]}$$

Ammonia, obtained by the Haber process at some expense, is combined with water to form the weak base, aqueous ammonia, NH$_3$(aq) or NH$_3$·H$_2$O.

$$\text{NH}_3 + \text{H}_2\text{O} \longrightarrow \text{NH}_3\text{·H}_2\text{O} \qquad \text{[27.3]}$$

At the same time the CO_2 produced according to equation [27.2] is added to water to form carbonic acid, a weak acid:

$$CO_2 + H_2O \longrightarrow H_2CO_3 \quad [27.4]$$

The weak base, $NH_3(aq)$, and weak acid, H_2CO_3, then undergo acid-base neutralization according to equation [27.5].

$$NH_3 \cdot H_2O + H_2CO_3 \longrightarrow NH_4HCO_3 + H_2O \qquad [27.5]$$

A saturated solution of NaCl is then added to the NH_4HCO_3 solution, resulting in precipitation of $NaHCO_3$.

$$NaCl + NH_4HCO_3 \longrightarrow NaHCO_3(s) + NH_4Cl \qquad [27.6]$$

Most of the $NaHCO_3$ produced commercially in this way is converted to Na_2CO_3 according to equation [27.1]. The NH_4Cl produced in reaction [27.6] is combined with $Ca(OH)_2$ ($CaO + H_2O$) to form $CaCl_2$, which has more demand commercially.

$$2\ NH_4Cl + Ca(OH)_2 \longrightarrow CaCl_2 + 2\ NH_3(g) + H_2O \qquad [27.7]$$

The NH_3 from reaction [27.7] and the CO_2 from reaction [27.1] are recycled to form more NH_4HCO_3. Thus $CaCl_2$ is the only by-product of the Solvay process.

In this experiment $NaHCO_3$ and Na_2CO_3 will be prepared according to the Solvay method with the modification that dry ice will be used as the source of the CO_2 instead of limestone. Also, NH_3 will be obtained from ammonia water (aqueous ammonia or ammonium hydroxide, NH_4OH). The sequence of additions will be further altered slightly.

SAFETY PRECAUTIONS

Review the safety rules on pages 1 and 2.

Handle concentrated aqueous NH_3 with extreme care. It can be very harmful if inhaled. Use in the hood.

Do not handle dry ice with your bare hands. It can cause burns. Use gloves.

Clean, rinse, and wipe any spills thoroughly.

PROCEDURE

Weigh about 15 g of sodium chloride to 0.01 g in a 250 mL beaker. Add 70 mL of concentrated aqueous NH_3 (15 M) in 5 mL portions with continuous stirring until the NaCl is dissolved. (Perform this step in the hood.) Add 50 mL of powdered dry ice (solid CO_2) quickly with continuous stirring until precipitation occurs.* After precipitation, add about 20 mL of powdered dry ice. The total volume of dry ice added should be about 70 mL. (The density of solid CO_2 is 1.5 g/mL.) Stir continuously until bubbling stops. Cool in an ice bath. Filter the precipitate on a suction filter and wash with one 3 mL portion of ice-cold water and one 3 mL portion of acetone. After suction drying for 5 min, transfer to an evaporating dish and dry in an oven at 110°C for at least 2 hours. After drying, weigh the product and determine the yield and percent yield.

---

*Alternatively, CO_2 gas can be bubbled through the solution from a CO_2 generator or compressed CO_2 cylinder.

Test of the Product

1. Transfer a small portion of the dried product to a watch glass. Add 1 or 2 drops of 3 M HCl. What happens? Compare this result with known samples of $NaHCO_3$ and Na_2CO_3.

2. Weigh about half of the remaining dried product in a crucible and heat carefully, gradually intensifying the heat over a 30 min period. Cool and weigh. Compare the mass of the product with that calculated according to equation [27.1].

3. Test the decomposed product from step 2 for carbonate ion by dissolving a small portion in 5 mL of water in a test tube and adding 5 mL of 1 M $CaCl_2$. If carbonate is present what should happen? Repeat the test on a sample of the undecomposed product. What conclusions can be drawn from these tests?

DISPOSAL

Solid NaHCO₃ and Na₂CO₃: Dispose in the laboratory trash container.

Excess ammonia solution: Dilute with an equal volume of water. Add phenolphthalein and neutralize with 6 M HCl or H_2SO_4 (first disappearance of permanent pink color). Flush the resulting salt solution down the sink with running water.

Report   Name _____ Section _____

                                                              Lab Partner _____

DATA

Mass of NaCl, g   _____

Mass of product, g   _____

Theoretical yield of NaHCO$_3$ _____

Percent yield   _____

HCl test (result)   _____

Mass of product (before heating), g   _____

Mass of product (after heating), g   _____

Theoretical yield of Na$_2$CO$_3$ _____

Percent yield   _____

Test for CO$_3{}^{2-}$ on product before heating (result)   _____

Test for CO$_3{}^{2-}$ on product after heating (result)   _____

Sample Calculations

Report   Name _____ Section _____

QUESTIONS

1. Did the reaction yield the expected sodium bicarbonate? Cite the experimental evidence you obtained to support your conclusion.

2. Did the reaction in step 2 of the test procedure yield the expected sodium carbonate? Cite the experimental evidence you obtained to support your conclusion.

3. How would each of the following errors affect the calculated percent yield of product (would the percent yield be high, be low, or not change)? Explain your answer.

   a. The sodium bicarbonate was not completely dry when weighed.

   b. Fourteen grams of NaCl were mistakenly used instead of the 15 grams called for in the procedure. Fifteen grams was recorded in the laboratory notebook.

   c. The sodium bicarbonate and crucible were not heated sufficiently to convert all of the sodium bicarbonate into sodium carbonate in step 2 of the test procedure.

Prelab   Name _____ Section _____

The Solvay Process: Preparation of NaHCO$_3$ and Na$_2$CO$_3$

1. Write chemical equations to describe the reactions taking place in the three procedure steps used to test the product.

2. What safety precaution should be observed when using concentrated ammonia?

3. What is the purpose of washing the precipitate with cold water and acetone? How will the actual yield be affected by washing?

4. Calculate the total number of moles of NaCl and CO$_2$ used in the procedure. Which reactant is the limiting reagent? Determine the theoretical yield of NaHCO$_3$.

5. Reaction [27-1] takes place in step 2 of the test procedure. What happens to the carbon dioxide and water that are produced in the crucible?

28 A Penny's Worth of Chemistry

INTRODUCTION

This experiment provides a means of becoming acquainted with some techniques of working carefully with metals in acid. The goal is to dissolve a copper disk in acid and, through a series of reactions, recover the copper. A schematic of the entire series of reactions follows, where the symbols over the arrows indicate concisely the reagents or techniques used in each step.

$$Cu(s) \xrightarrow{HNO_3} Cu^{2+}(aq) \xrightarrow{NaOH} Cu(OH)_2(s) \xrightarrow{\Delta} CuO(s) \xrightarrow{H_2SO_4} Cu^{2+}(aq) \xrightarrow{Zn} Cu(s)$$

Copper dissolves in either concentrated nitric acid (equations [28.1, 28.2 and 28.3]) or in dilute nitric acid (equations [28.4, 28.5 and 28.6]). Each of these chemical reactions can be described by three chemical equations: a molecular equation, a full ionic equation and a net ionic equation. The molecular equation gives the formula of the reactants that are added together and treats all products as if they exist as intact compounds.

Copper dissolving in concentrated nitric acid

$$Cu(s) + 4\ HNO_3(aq) \longrightarrow Cu(NO_3)_2(aq) + 2\ NO_2(g) + 2\ H_2O \qquad [28.1]$$

The molecular equation can be used to write the full ionic equation in which all aqueous strong electrolytes (strong acids, strong bases, and soluble ionic compounds) are broken into their respective aqueous ions

$$Cu(s) + 4\ H^+(aq) + 4\ NO_3^-(aq) \longrightarrow Cu^{2+}(aq) + 2\ NO_3^-(aq) + 2\ NO_2(g) + 2\ H_2O \qquad [28.2]$$

For any substance that appears on both sides of the chemical equation, an equal number of moles of that substance can be removed from the chemical equation. In the full ionic equation there are four moles of nitrate on the reactant side of the equation and two moles on the product side. Therefore, two moles of nitrate can be removed from both sides of the equation to yield that net ionic equation.

$$Cu(s) + 4\ H^+(aq) + 2\ NO_3^-(aq) \longrightarrow Cu^{2+}(aq) + 2\ NO_2(g) + 2\ H_2O \qquad [28.3]$$

Copper dissolving in dilute nitric acid

Molecular equation

$$3\ Cu(s) + 8\ HNO_3(aq) \longrightarrow 3\ Cu(NO_3)_2(aq) + 2\ NO(g) + 4\ H_2O \qquad [28.4]$$

Full ionic equation

$$3\ Cu(s) + 8\ H^+(aq) + 8\ NO_3^-(aq) \longrightarrow 3\ Cu^{2+}(aq) + 6\ NO_3^-(aq) + 2\ NO(g) + 4\ H_2O \qquad [28.5]$$

Net ionic equation

$$3\ Cu(s) + 8\ H^+(aq) + 2\ NO_3^-(aq) \longrightarrow 3\ Cu^{2+}(aq) + 2\ NO(g) + 4\ H_2O \qquad [28.6]$$

The two reactions can be distinguished from one another since $NO_2(g)$ has a red-brown color and $NO(g)$ is colorless. Two equilibria confuse the issue, however, as represented by equations [28.7] and [28.8].

$$2\ NO(g) + O_2(g) \rightleftharpoons 2\ NO_2(g) \qquad K_{eq} = 2.32 \times 10^{12} \qquad [28.7]$$

$$2\ NO_2(g) \leftrightharpoons N_2O_4(g) \qquad K_{eq} = 6.86 \qquad\qquad\qquad [28.8]$$

$N_2O_4(g)$ is also colorless. Because of these complications, most reactions of nitric acid with copper look similar. The reaction with concentrated nitric acid is much more rapid, however, and it is this one that we shall use.

Note: $NO_2(g)$ is a deadly poison! To quote from a reliable source[*]: **"One of the most insidious gases. Inflammation of lungs may cause only slight pain or pass unnoticed, but the resulting edema several days later may cause death. 100 ppm is dangerous for even a short exposure, and 200 ppm is fatal."** In a laboratory of average size, 100 ppm = 18.6 g = 0.404 mol = the amount of NO_2 produced by dissolving 12.8 g Cu, or approximately four copper disks. It is therefore essential that you work in the hood when dissolving the disk in nitric acid. This warning is included so that you can protect yourself if some NO_2 is released in the laboratory. **If you notice NO_2 fumes outside the hood, do not stand around. Leave the laboratory immediately!**

We would find it very difficult to recover the copper from this strongly oxidizing acidic nitrate solution, since any elemental copper produced will be oxidized immediately by the excess nitrate ion present according to equations [28.4] through [28.6]. This oxidation of copper metal is catalyzed by the presence of acid. Accordingly, we shall remove as much as possible both the H^+ and NO_3^- from the solution.

Addition of sodium hydroxide solution to the blue acidic copper(II) nitrate solution causes two things to happen. First, the hydrogen ion in solution reacts with the added hydroxide ion to form water. The hydrogen ion is effectively removed from the solution.

$$H^+(aq) + OH^-(aq) \longrightarrow H_2O$$

Second, the copper(II) ion forms an insoluble substance with the hydroxide ion, copper(II) hydroxide. The copper(II) hydroxide produced is difficult to work with since it is a rather gelatinous white precipitate (with some traces of blue). Therefore, we heat the solution to produce black copper(II) oxide.

This still leaves the nitrate ion in solution, but this is removed by decanting the liquid. You want to allow the black copper(II) oxide to settle to the bottom and then carefully pour out the supernatant solution–the solution on top of or above a solid–making sure not to lose any of the solid. Some nitrate ion will still be present, so we add some hot water, stir, allow the copper(II) oxide to settle, and then decant the liquid again. This process is called "washing" the precipitate—in this case the copper(II) oxide. Since you will need about 150 mL of almost boiling water to do this, you can save considerable time if you put the water on to boil at the beginning of the laboratory period.

The copper is now on the bottom of the beaker in the form of copper(II) oxide, and we would like to have it in solution. This is done readily by adding sulfuric acid. The copper(II) oxide, which is basic, reacts with sulfuric acid to produce a copper(II) sulfate solution with a striking blue color.

Since sulfuric acid will not dissolve copper (try it with a piece of copper wire if you are skeptical), we can now recover the copper from the solution by reducing it with zinc. This is accomplished by adding fine particles of zinc to the solution. Of course, zinc also reacts with the excess acid to produce hydrogen, so large quantities of gas are evolved on the addition of zinc. Be careful that the hydrogen produced is not released into the laboratory because it can explode. This reaction must be performed under the hood. The excess zinc is removed by adding some hydrochloric acid.

The only solid now left is copper, which must be dried so that it can be weighed. If you do this over a Bunsen burner flame, the copper will react with the oxygen in the air to produce black copper(II) oxide. Consequently, wash the copper first with methanol to remove the water and then with acetone, which will rapidly evaporate. The acetone is removed by evaporation over a hot plate or over a steam bath (Figure 28-2).

---

[*]Martha Windholtz, ed., *The Merck Index*, 10th ed., Gahway, NJ: Merck and Co., 1983, p. 947.

SAFETY PRECAUTIONS

Review the safety rules on pages 1 and 2.

Work in the hood when dissolving the copper disk with nitric acid, and when adding granular zinc or hydrochloric acid to the mixture.

Keep methanol and acetone away from all open flames. Both substances are extremely flammable.

Add all reagents slowly, because rapid addition may cause an unexpectedly fast reaction that may spatter.

Stir the copper hydroxide solution constantly while it is being heated to prevent spattering.

Use insulated gloves, or beaker tongs (not crucible tongs), to handle vessels containing hot water or hot solutions, but be careful that the vessel does not slip out of the gloves or tongs and break. Practice first with a 400 mL beaker filled with cold water.

PROCEDURE*

1. Working in a group of four weigh a copper disk to the nearest milligram. Place the disk in a 400 mL beaker.

2. Place 150 mL of distilled water in a 250 mL beaker. Place the beaker on a wire gauze on a ring stand, and heat with a bunsen burner until the water boils. Do not wait for the water to boil but proceed with the rest of the experiment. You will use this hot water in step 6.

3. In the hood add 24.0 mL of concentrated nitric acid to the copper disk in the beaker. When the reaction is complete (the copper disk is gone and no gas bubbles are being produced) add 70 mL of water. You may now take the beaker out of the hood. Be careful. The beaker may be hot! When the beaker has cooled so that it may be handled safely, pour its contents into a 100 mL graduated cylinder. Rinse the beaker with two 5 mL portions of distilled water and add this to the graduated cylinder. Add distilled water to bring the total volume to 100 mL. Carefully stir the solution with a glass stirring rod to assure that the solution is homogeneous. Measure 25.0 mL of solution into 100 mL beakers for each student in the group, rinsing the graduated cylinder with distilled water into the last beaker. The remainder of the experiment will be done on an individual basis using the 25 mL of solution. This solution corresponds to one fourth of the copper disk.

4. Add 45 mL of 3.0 M sodium hydroxide solution to form copper(II) hydroxide. Stir well.

5. Add two or three boiling chips. Then, with stirring, heat the solution just to boiling to convert the copper(II) hydroxide to copper(II) oxide. Be very careful to heat gradually and just until the black copper(II) oxide is formed from the blue-white copper(II) hydroxide. Too much heat will cause the solution to bump and spatter all over you. If you do not add the boiling chips, your solution will "bump" when you heat it and spatter out of the beaker. The boiling chips prevent this by giving the vapor bubbles a place to form. Place the beaker on the bench top. Use beaker tongs (not crucible tongs) or insulated gloves to move the hot beaker.

6. Allow the solid to settle and then decant the supernatant liquid into another beaker. It is sufficient to let the solid settle for 2 to 3 min or until about a 2 cm depth of relatively clear liquid appears at the top of the mixture. To decant means to carefully pour off the liquid making sure that none of the solid goes with it. Add about one-third (50 mL) of the very hot water you prepared, stir up the solid for about 15 sec, allow the solid to settle, and decant the hot water. Repeat this with two more portions of very hot water. Handle the beakers with insulated gloves or beaker tongs.
   ### DISPOSAL

   Decanted solutions: Add phenolphthalein and, if basic, neutralize with 6 M H_2SO_4 (first disappearance of pink color). Flush this salt solution down the sink with running water.

---

* George F. Condike, *J. Chem. Educ.*, **52**, 1975, p. 615.

7. Add 25 mL of 6.0 M H_2SO_4 solution to produce copper(II) sulfate. Stir this mixture to get it to react completely. It will also get somewhat warm and you would be wise to hold the beaker with tongs or insulated gloves. Make sure that all the black copper(II) oxide dissolves. You may have to scrape some of this black solid down from the top edges of the beaker with a wet stirring rod. Decant the solution into another beaker and discard the boiling chips.

8. In the hood, add 3.0 g of 30-mesh zinc metal slowly to precipitate copper metal. Stop adding if the reaction foams a lot. Wait until the foaming stops, then resume adding zinc. Stir until the supernatant solution has become colorless. If the reaction ceases and the solution remains blue, add a small additional amount of zinc. When no blue color is present and you can no longer observe gas being formed, decant the supernatant liquid into another beaker. Discard the liquid once you are sure no solid has been transferred.

    DISPOSAL

    Acid–zinc(II) solution: Mix and save for combining with waste from step 11.

9. Still in the hood, add 10.0 mL of distilled water and then 15.0 mL of concentrated hydrochloric acid to destroy the excess zinc. When the evolution of gas bubbles has become very slow, warm, but do not boil, the mixture. Do not warm the mixture too much. A maximum temperature of about 60°C is what you are trying to attain.

10. While you are waiting for the gas to stop evolving, weigh a porcelain evaporating dish to the nearest mg. When you can no longer see any gas bubbles forming, add 50 ml of distilled water, remove from the hood, and stir. Then decant and discard the supernatant liquid into another beaker and transfer all the solid (use a glass rod) to the weighed evaporating dish. You may use some distilled water to help in transferring the solid.

    DISPOSAL

    Acid–zinc(II) solution: Mix and save for combining with waste from step 11.

11. Add 10.0 mL total of distilled water to the solid, stir the mixture to make sure that all the solid is washed, and decant the distilled water into a beaker. Repeat this washing with two more 10.0 mL portions of distilled water. Discard the decanted distilled water.

    DISPOSAL

    Acid–zinc(II) solution: Combine with the waste solutions from steps 8 and 10. Add phenolphthalein and neutralize with 6 M NaOH (first appearance of pink color). Dispose in a waste bottle labeled *heavy metals—solutions.*

12. Wash the solid with 10 mL of methanol. (This substance is highly flammable. Keep it away from flames!) Decant the liquid into a beaker, and discard the washings.

    DISPOSAL

    Methanol/acetone: *Dispose* in a waste bottle labeled *organic–aliphatic–nonhalogenated* or labeled with the compound names.

13. Wash the solid with 10 mL of acetone. (This substance is highly flammable. Keep it away from flames!) Decant the liquid into a beaker, and discard the washings.

    DISPOSAL

    Methanol/acetone: Follow instructions for disposal in step 12.

14. Evaporate the remaining acetone by placing the evaporating dish on a hot plate or on a steam bath as illustrated in Figure 28-1. Ensure that the evaporation is as complete as possible by stirring the solid

with a glass rod. When dry, the fine particles of copper metal will be granular and will no longer adhere to the stirring rod.

15. Dry as much of the evaporating dish as you can (don't forget the bottom). You can use paper towels to dry the evaporating dish. Do not get overzealous and try to dry the copper, for some of it will stick to the paper towels and be lost. Wait for the evaporating dish to cool and weigh it with your product to the nearest milligram.

16. If you have enough time, repeat steps 14 and 15 until two successive weights are the same. Good chemists always perform this final step to make sure that their product is completely dry.

DISPOSAL

Copper metal: Put in a recycling bottle labeled *copper metal*.

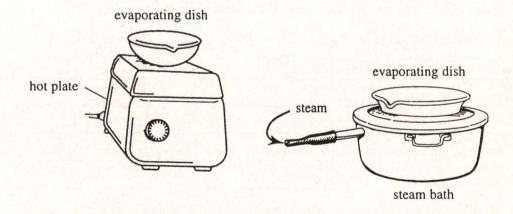

Figure 28-1. Two Methods for Evaporating Off Acetone

Report   Name _____ Section _____

DATA

_____

	Trial 1	Trial 2
Mass of disk and 400 mL beaker, g	_____	_____
Mass of 400 mL beaker, g	_____	_____
Mass of disk, g	_____	_____
Average mass of disk, g	_____	
Mass of product and evaporating dish, g	_____	_____
Mass of evaporating dish, g	_____	_____
Mass of product, g	_____	_____
Average mass of product, g	_____	
Percent recovery (based on whole copper disk)	_____	

For each of the following steps, give the color and formula of the indicated substance produced in this step. For solution formulas, you may give the formula of the ion that produces the color.

Step	Substance	Color	Formula
3	solution	_____	_____
3	gas	_____	_____
4	solution	_____	_____
4	solid	_____	_____
5	solution	_____	_____
5	solid	_____	_____
7	solution	_____	_____
8	solution	_____	_____
8	gas	_____	_____
8	solid	_____	_____
9	solution	_____	_____
9	gas	_____	_____

Report   Name _____ Section _____

QUESTIONS

1. (a) Write the full balanced molecular equation and the balanced net ionic equation for the reaction that has occurred in each of the following steps. Indicate the state of matter of each species in the balanced equations. (b) Classify the reaction by type (combination, decomposition, single displacement, metathesis, or redox). Note that more than one classification type may fit– for example, a reaction may be single displacement and redox. If so, indicate both types. (c) State what leads you to conclude that a reaction has occurred (evolution of a gas, change of color, heat, light, sound, formation of a precipitate, or any combination).

Step 3

Molecular equation _____

Net ionic equation _____

Classification _____

Observations _____

Step 4

Molecular equation _____

Net ionic equation _____

Classification _____

Observations _____

Step 5

Molecular equation _____

Net ionic equation _____

Classification _____

Observations _____

Step 7

Molecular equation _____

Net ionic equation _____

Classification _____

Observations _____

Report   Name _____ Section _____

   Step 8 (two reactions)

   Molecular equation   _____

   Net ionic equation   _____

   Classification   _____

   Observations   _____

   Molecular equation   _____

   Net ionic equation   _____

   Classification   _____

   Observations   _____

   Step 9

   Molecular equation   _____

   Net ionic equation   _____

   Classification   _____

   Observations   _____

2. Was your percent recovery of copper greater than or less than 100%? Describe an experimental error
    that could have caused this. What could be done to prevent this error?

Prelab   Name _____ Section _____

A Penny's Worth of Chemistry

1.   What is the safety precaution associated with dissolving a copper disk?

2.   What are the name and chemical formula of the product of the reaction in step 4 of the procedure?

3.   What are the name and chemical formula of the gaseous product of the reaction in step 8 of the procedure?

4.   What are the name and chemical formula of the solid product of the reaction in step 8 of the procedure?

5. What is meant by the phrase "washing the precipitate"?

29

Polymer Synthesis

INTRODUCTION

Polymers are large molecules made of many (poly-) small units. The starting material, which is a single unit, is called a *monomer*. Many of the most important biological compounds are polymers. Cellulose and starch are polymers of glucose units, proteins are made up of amino acids, and nucleic acids are made up of nucleotides. Since the 1930s a large number of synthetic polymers have been manufactured. Synthetic fibers such as nylon and polyesters, polyethylene and polystyrene used in packaging materials, and polyvinyl chloride have become common in our everyday lives.

This experiment will focus on man-made polymers and the basic mechanism by which some of them are produced. Two of the most important types of reaction used in polymer formation are the addition and the condensation polymerization reactions. The polymerization of styrene is formed by the addition reaction. Nylon is synthesized by the condensation polymerization reaction.

Addition Polymers

One of the basic types of polymers is that formed by the addition of units originally having double bonds. For example, styrene is an organic monomeric unit that contains a double bond and can undergo addition polymerization.

$$H_2C=CH(C_6H_5) + H_2C=CH(C_6H_5) \longrightarrow H_3C-CH(C_6H_5)-CH=CH(C_6H_5) \qquad [29.1]$$

Since two monomer units are added to each other with the elimination of the double bond it is referred to as an *addition reaction*. The reaction does not begin to take place without the help of a catalyst called an *initiator*, which starts the reaction. The initiators commonly used in these reactions are peroxides, such as benzoyl peroxide. In the presence of heat or UV light, the benzoyl peroxide splits into two halves producing two free radicals. The double bond of the benzoyl peroxide splits yielding two species each of which has one of the electrons from the shared double bond. The resulting species that has one unpaired electron is called a *free radical*. The boldfaced dot represents the unpaired electron. The free radical reacts with the sytrene and initiates the reaction.

$$C_6H_5COO\bullet + H_2C=CH(C_6H_5) \longrightarrow C_6H_5COOCH_2CH(C_6H_5)\bullet \qquad [29.2]$$

We now have one molecule, twice as large as the first. The reaction will continue adding a styrene monomer to the growing chain until a giant molecule is formed containing many thousands of styrene-repeating units.

$$CH_3CH \ C_6H_5 \ [CH_2CH \ C_6H_5]_n CH_2CH_2 \ C_6H_5$$

The polymer Lucite, commonly known as Plexiglas, is another example of a polymer formed by the addition reaction. The monomeric units of the polymer are methyl methacrylate. The reaction is catalyzed by benzoyl peroxide.

$$n \ H_2C=C(CH_3)COOCH_3 \longrightarrow \ -(CH_2C(CH_3)COOCH_3)_n \qquad [29.3]$$

Condensation Polymers

In condensation reactions, the monomer contains a reactive group other than the double bond. These monomers react by splitting out a small molecule such as water from between the two reactive functional groups. Two common reactions of this type are the reactions of organic acids with alcohols and amines to give either esters or amides. In order for this reaction to continue, each monomeric unit must have two reactive functional groups. The dimer (product of the first condensation) is capable reacting further because it still has a functional group free to react. For example, the condensation of adipoyl chloride and hexamethylenediamine will produce the polyamide polymer known as Nylon 6-6.

$$ClCOCH_2CH_2CH_2CH_2COCl + H_2NCH_2CH_2CH_2CH_2CH_2CH_2NH_2 \longrightarrow$$
adipoyl chloride   *hexamethylenediamine*

$$ClCO(CH_2)_4CO[NH(CH_2)_6NHCO((CH_2)_4CO]NH(CH_2)_6NH_2 + HCl \qquad [29.4]$$

An amide link is formed between the adipoly chloride and the amine and HCl is eliminated. The polymer is called nylon 6-6 because there are six carbon atoms in the adipoyl chloride and six carbon atoms in the hexamethylenediamine. There are several forms of nylon, differing in the length of the carbon chains in each of the components. In this and other chemical reactions involving acids, a derivative of the acid–not the acid itself–is used. These acid derivatives are much more reactive than the acids themselves. In this reaction, NaOH is added to the polymerization reaction to neutralize the HCl formed during the condensation reaction that forms the amide link. Polyesters are formed when the acid group reacts with an alcohol (hydroxyl) group to produce an ester. Again there must be at least two reactive groups.

In both types of polymerization reaction, the length of the polymer chain is dependent on the environmental conditions. Usually the chains can be made longer by heating the products further. This process is referred to as *curing*. It is difficult to write a specific reaction for a polymerization since we do not know how many monomeric units have condensed to form the polymer. The letter *n* is used to designate an indefinite number of the repeating unit. The reaction for the formation of a polyester is written as follows:

$$n\ HOOCCH_2COOH + n\ HO(CH_2)_2OH \longrightarrow n\ HOOCCH_2COOH(CH_2)_2OH + n\ H_2O \qquad [29.5]$$

SAFETY PRECAUTIONS

Review the safety rules on pages 1 and 2.

The initiators are organic peroxides. These substances are hazardous. They are irritating to the skin, eyes, and mucous membranes. They should not be ground or subjected to heat or shock, since explosive decomposition may occur. They should NOT be added to HOT material. They should be kept away from heat, sparks, and open flames. They should be stored in a cool, dry place away from sunlight. Do not breathe the vapors given off by the pure chemicals or the mixtures involved in this experiment. Keep all beakers covered with watch glasses. **All work should be done in the hoods.**

Wash your hands thoroughly with soap and running water after you have completed this experiment.

PROCEDURE

A. Preparation of Addition Polymers

1. In the hood, pour 10 mL of the casting resin (styrene) into a 50 mL beaker. Note the time, and record it in your notebook.

2. Add approximately 8 drops of the initiator catalyst, methyl ethyl ketone peroxide (t-butyl peroxide benzoate can be substituted if necessary). Stir slowly with a glass stirring rod. The polymerization reaction is exothermic and therefore generates its own heat. Mix thoroughly without introducing any air bubbles. Clean the stirring rod when mixing is complete.

3. Let the mixture stand at room temperature for 10 min. It will become viscous on standing. If you wish at this time, you can imbed an object such as a coin or a marble by carefully positioning the object and allowing it to sink to the bottom of the beaker. Let the mixture sit for about an hour in order for the solution to thoroughly solidify.

4. At the end of an hour, the polymer should have shrunk sufficiently to remove it from the beaker. You may keep the product. Examine the object and note its physical characteristics.

   An alternative procedure is to pour the contents of the beaker onto a watch glass and let it solidify. The residual polystyrene in the beaker can be removed by rinsing the beaker with a few milliliters of xylene and warming it on a hot plate until the polymer is dissolved. Pour a few drops of the warm xylene solution on a microscope slide and let the solvent evaporate under the hood. A thin film of polystyrene will be left on the slide. This is a technique referred to as *solvent-casting* that is used to make films from bulk polystyrene. The remaining xylene mixture can be discarded into an appropriate waste container. The beaker can be washed with soap and water. The solid polystyrene on the watch glass can be pried off with a spatula. Observe its consistency and characterize is physical properties.

B. Preparation of Condensation Polymers (Nylon)

1. Add 2.0 mL of 20% NaOH solution and 10 mL of 5% aqueous solution of hexamethylenediamine to a 50 mL reaction beaker.

2. Draw 10 mL of 0.25 M adipoyl chloride solution in cyclohexane into a pipet. Layer this solution carefully and slowly on top of the aqueous layer in the beaker. At the end of the process there should be two distinct layers in the beaker. Nylon will form at the interface of the two layers.

3. Place the beaker on a pad of paper towels. Insert a spatula, pair of tweezers, or wire bent into the shape of a hook to the film layer between the two liquids. Then slowly lift the film from the center. Pull is slowly. If you pull it too fast the nylon thread will break.

4. Gently wind the nylon thread around the spatula. Tip the beaker slightly and continue to pull out the film. Guide the thread of nylon over the spout of the beaker onto the paper towel pad. If the thread breaks, you can pick it up again at the interface between the two solutions. The process can be repeated until one of the solutions is used up.

5. If you want to keep the nylon thread, wash it thoroughly with water and let it dry on a clean paper towel. Do not handle the nylon until it has been thoroughly washed.

6. Dissolve a small amount of nylon in 80% formic acid. Place a few drops of the solution onto a microscope slide and allow the solvent to evaporate under the hood. Compare the appearance of the solvent cast nylon film with that of the nylon thread.

DISPOSAL

Dispose of excess reagents in the containers provided. *Do not pour them down the sink.*

Report   Name _____ Section _____

1. Describe the appearance of the polystyrene and nylon.

2. Could you distinguish between polystyrene and nylon on the basis of solubility?

3. Is there any difference in the appearance of the solvent cast films of nylon and polystyrene?

Report   Name _____ Section _____

QUESTIONS

1. A polyester is made of sebacoyl chloride, $ClOC(CH_2)_8COCl$, and ethylene glycol, $(OH)CH_2CH_2(OH)$.

    a. Draw the structure of the polyester formed from these two monomers.

    b. What molecules have been eliminated in this condensation reaction?

2. Dacron is a polyester made of terephthalic acid and ethylene glycol

$$C_6H_5(COOH)_2 + (OH)CH_2CH_2(OH) \longrightarrow$$

Draw the structure of Dacron. What molecule has been eliminated?

Prelab   Name _____ Section _____

Polymer Synthesis

1. Distinguish between a monomer and a repeating unit.

2. Give an example of a polymer in each of the following classes:
   a. A condensation polymer

   b. An addition polymer

   c. A polyester

3. Write the reaction for the polymerization of ethene.

4. If nylon were formed with a dicarboxylic acid and a diamine, what would be the elimination product?

5. What is the safety precaution associated with the initiators used in this experiment?

30 Preparation of [Co(NH₃)₅ONO]Cl₂-Linkage Isomerism

INTRODUCTION

One of the interesting aspects of coordination compounds is the possibility of the existence of the compound in various isomeric forms. Isomeric possibilities include stereoisomers (both optical and cis-trans) as well as structural isomers. Although stereoisomers differ only minimally (in certain properties), structural isomers often differ markedly in many respects. Structural isomers are different, depending upon the nature of the ligand within the coordination sphere compared with others outside the sphere. For example, consider the following pairs,

$$[Cr(H_2O)_6]Cl_3 \text{ and } [Cr(H_2O)_4Cl_2]Cl \cdot 2 H_2O \qquad [30.1]$$

$$[Co(NH_3)_5Br]SO_4 \text{ and } [Co(NH_3)_5SO_4]Br \qquad [30.2]$$

In hydrate isomerism [30.1], the chloride ion and water can replace each other in the coordination sphere (in brackets), while in ionization isomerism, [30.2] the bromide and sulfate ions interchange with each other.

Linkage isomers differ because of the possibility of more than one donor atom on the ligand. The nitrite ion, NO_2^-, is one such ligand. When this ion coordinates with a metal atom using nitrogen as the donor atom, a nitro ($-NO_2$) compound is formed. If instead coordination takes place through oxygen as the donor atom, a nitrito ($-ONO$) compound results. Depending upon the reaction conditions during synthesis, one isomer can be favored, even though the other isomer might be more stable thermodynamically.

In this experiment pentaamminenitritocobalt(III) chloride, $[Co(NH_3)_5ONO]Cl_2$, will be prepared. Upon standing or in light or heat, this compound isomerizes to the more stable nitro form, pentaamminenitrocobalt(III) chloride, $[Co(NH_3)_5NO_2]Cl_2$. A subtle but distinct color change occurs during the isomerization from salmon-pink to yellow-brown. Such linkage isomers also differ slightly in other properties such as solubility and electronic and infrared spectra, and these differences may be used to further characterize the two isomers. See your text or the following references for further details on these and isomers of other coordination compounds:

J.P. FAUST and J.V. QUAGLIANO, *J. Am. Chem. Soc.*, **76** (1954): 5346.

J.L. BURMEISTER and F. BASOLO, *Inorg. Chem.*, **3**(1964): 1587.

G. PASS and H. SUTCLIFFE, *Practical Inorganic Chemistry,* London: Chapman & Hall 1968.

SAFETY PRECAUTIONS

Review the safety rules on pages 1 and 2.

Be particularly careful in handling both the NH$_3$ and H$_2$O$_2$. They can be quite harmful.

Clean, rinse, and wipe any spills thoroughly.

PROCEDURE

In a 250 mL Erlenmeyer flask, dissolve 5.0 g of ammonium chloride in 30 mL of concentrated aqueous ammonia (use hood). While stirring the solution continuously (magnetically), add 10 g of finely powdered hexaaquacobalt(II) chloride, [Co(H$_2$O)$_6$]Cl$_2$, in small portions. Continue to stir the resulting brown slurry while slowly adding 8 mL of 30% hydrogen peroxide from a dropping funnel. (*Caution*: Use extreme care in handling the 30% H$_2$O$_2$; it is quite corrosive.) When effervescence has ceased, filter any slight precipitate that might remain and cool the filtrate to about 10°C. While cooling, add 3 M HCl dropwise with a medicine dropper pipet until it is just neutral to litmus. Then add 5.0 g of sodium nitrite to the solution while stirring to dissolve. Allow the solution to stand in an ice bath for about 35 to 40 min, then filter the precipitated crystals of [Co(NH$_3$)$_5$ONO]Cl$_2$. Wash with 20 mL of icewater, followed by 20 mL of alcohol, and dry at room temperature. Weigh to determine the yield and percent yield. Note the color of your product immediately after preparation and again after standing for several days.

DISPOSAL

Solid product: Dispose in a waste bottle labeled *heavy metals—solids*.
Wash solution: Dispose in a waste bottle labeled *methanol wastes*.

Report

DATA

Limiting reagent _____

Mass of limiting reagent _____

Theoretical yield _____

Actual yield _____

Percent yield _____

Color of product _____

Sample Calculations

Report   Name _____ Section _____

QUESTIONS

1. Did your synthesis produce the desired product? Cite the evidence to support your answer.

2. Write the balanced equations for the reactions occurring during this preparation.

3. Which isomer is more stable? Cite *experimental* evidence to support your answer.

4. Given that [Co(H$_2$O)$_6$]Cl$_2$ is the limiting reagent, determine the theoretical yield of the product.

5. Will the mass of product increase, decrease, or stay the same if the color changes from salmon-pink to yellow-brown? Explain.

Prelab   Name _____ Section _____

Preparation of [Co(NH₃)₅ONO]Cl₂—Linkage Isomerism

1. Define *linkage isomers*.

2. What is the coordination number and oxidation state of the cobalt in the product compound? Show a calculation to support your answer for the oxidation state.

3. Draw structures for the two isomeric compounds being studied in this experiment. Clearly show the difference between the two structures.

4. What is the function of the H_2O_2 in this experiment? What is the gas being given off during the effervescence of H_2O_2?

5. Why is the product dried at room temperature rather than in an oven?

31 Free Radical Bromination of Organic Compounds

INTRODUCTION

The relative rates at which bromine reacts with a series of organic hydrocarbons will be determined. Under the conditions employed in this experiment, a bromine atom is substituted for a hydrogen atom in the hydrocarbon. For example, in the presence of bromine, ethane yields bromoethane and hydrogen bromide.

$$CH_3-CH_3 + Br_2 \rightarrow CH_3-CH_2-Br + HBr$$

ethane   bromoethane

The compounds used in this experiment are quite similar in that they contain only carbon and hydrogen. Yet the relative rates at which bromination occurs cover a wide range. The rate of the reaction depends on the exact structure of the hydrocarbon involved. For example, the presence of a benzene ring in a hydrocarbon causes a dramatic change in the rate of the reaction. Yet some exceptions will be observed that will need to be explained. The questions in the report will help you draw some very specific conclusions about the interaction of bromine with organic compounds.

SAFETY PRECAUTIONS

Review the safety rules on pages 1 and 2.

Use extreme care when dealing with carbon tetrachloride, a toxic compound that has been listed as a carcinogen by the U. S. Environmental Protection Agency. *This entire experiment must be conducted in an efficient hood.*

Do not breathe the vapors given off by the pure chemicals or the mixtures involved in this experiment. Keep all beakers covered with watch glasses.

Bromine is highly toxic through inhalation, ingestion, or skin contact. *Work with care in all phases of this experiment.*

Wash your hands thoroughly with soap and running water after you have completed this experiment.

Do not dispose of reaction mixtures down the drain. Discard them in a designated container in the hood.

All glassware used for carbon tetrachloride solutions must remain in the hood.

PROCEDURE — WORK IN THE HOOD!!

Place the assigned volume of the hydrocarbon and 30 mL of carbon tetrachloride in a 150 mL beaker and cover it with a watch glass. Place the beaker on a sheet of white paper. In a 50 mL graduated beaker, place 25 mL of a 2% (mass/volume) solution of bromine in carbon tetrachloride (Br_2/CCl_4) and cover with another watch glass. **Caution: Avoid breathing the fumes associated with this experiment. Always keep both beakers covered with watch glasses.** To start the reaction, rapidly pour the bromine solution into the 150 mL beaker containing the hydrocarbon. *Immediately replace the watch glasses.* Determine the time required for the reddish color of bromine to disappear and record this time interval. If no color change occurs, stop after one hour has elapsed. If the color of the bromine disappears, lift the watch glass covering the beaker slightly. Visually observe (do not smell) the fumes that should be present. Moisten a piece of blue litmus paper and support it at the top of the beaker between the watch glass and the lip of the beaker. Record any observations.

Upon completion of the reaction, the resulting solution may be discarded into a large container with sodium hydroxide solution. The beakers used to dispense the bromine solution and the reaction beaker itself should *always* remain in the hood.

Run the reaction with one or more of the following compounds as assigned by your instructor: benzene, toluene, ethylbenzene, cumene, xylene, *t*-butylbenzene, *n*-propylbenzene or *n*-butylbenzene, pentane or hexane, cyclohexane, cyclohexene (optional), and cyclooctatetraene (optional). Each compound should be run twice. Class data will be collected and pooled. Record these data in your laboratory notebook in tabular form.

DISPOSAL

All mixtures: Dispose in a waste bottle labeled *organic—aromatic—halogenated.*

Report

Name _____ Section _____

Name and Formula of Compound	Molecular Mass	Assigned Volume, mL	Density g/mL	Mass, g	Amount, mol	Time min: sec	Time (av), sec
Benzene C_6H_6	_____	6.0	0.8787	_____	_____	_____	_____
Toluene C_7H_8	_____	7.0	0.8669	_____	_____	_____	_____
Ethylbenzene C_8H_{10}	_____	8.0	0.8670	_____	_____	_____	_____
Xylene C_8H_{10}	_____	8.0	0.8642	_____	_____	_____	_____
Cumene C_9H_{12}	_____	9.0	0.8618	_____	_____	_____	_____
n-Propylbenzene C_9H_{12}	_____	9.0	0.8620	_____	_____	_____	_____
t-Butylbenzene $C_{10}H_{14}$	_____	10.0	0.8665	_____	_____	_____	_____
n-Butylbenzene $C_{10}H_{14}$	_____	10.0	0.8601	_____	_____	_____	_____
Hexane C_6H_{14}	_____	8.5	0.6603	_____	_____	_____	_____
Cyclohexane C_6H_{12}	_____	7.0	0.7786	_____	_____	_____	_____
Cyclohexene C_6H_{10}	_____	6.5	0.8102	_____	_____	_____	_____
Cyclooctatetraene C_8H_8	_____	7.5	0.9206	_____	_____	_____	_____

Sample Calculations

Report   Name _____ Section _____

QUESTIONS

1. List the compounds used in the experiment in the order of most reactive (first) to least reactive (last).

2. In most cases, how does the presence of a benzene ring in a molecule affect the rate at which the substitution reaction with bromine takes place? What is (are) the exception(s)?

3. In general, does it appear that substitution of the bromine atom for a hydrogen atom in an alkyl substituted benzene most likely takes place on the benzene ring itself or on the side chain? Explain your reasoning.

4. Compare the rates of reaction of the alkyl substituted benzene. Are the rates directly related to the total number of alkyl (that is, nonaromatic) hydrogens present? Explain your reasoning.

5. Which specific carbon (in relation to the benzene ring) in the alkyl side chain most likely has a hydrogen on it replaced with a bromine? How do the experimental data for *t*-butylbenzene support your conclusion?

6. Should the rate of bromination for xylene be approximately twice that for toluene? Explain your answer.

Report
Name _____ Section _____

7. Complete the following statement: "When the rates of toluene, ethylbenzene, and cumene are compared, the more hydrogen atoms there are on the carbon of the side chain that is attached directly to the benzene ring, the _____ (faster or slower) the reaction goes."

8. Should the rate of bromination of ethylbenzene be approximately the same, twice, or half that for *n*-butylbenzene? Explain.

9. Benzene is unreactive toward Br_2/CCl_4. Is any cyclic compound that contains double bonds or delocalized π bonding unreactive to this reagent? How do the data support your answer?

10. Why were the volumes of the hydrocarbons varied– for example, why wasn't 6 mL of hydrocarbon used in each reaction, while the volume of Br_2/CCl_4 was kept constant?

11. What is the limiting reagent in these reactions, bromine or the hydrocarbon? Show calculations for at least three cases.

12. Draw the structures for the product of the bromination of each of the following compounds: toluene, ethylbenzene, cumene, xylene, *t*-butylbenzene, *n*-propylbenzene, and cyclohexene.

Prelab   Name _____ Section _____

Free Radical Bromination of Organic Compounds

1. Draw the structure of each of the following compounds below its name.

   benzene   toluene   ethylbenzene

   n-butylbenzene   cyclohexane   t-butylbenzene

   n-propylbenzene   *n*-pentane   cumene

   xylene   *n*-hexane   cyclohexene

   cyclooctatetraene

2. Which of the above compounds are aromatic?

3. Which of the above compounds are alkenes?

4. If the rate of a reaction is fast, will the reaction take a long or short time to occur? Explain your reasoning.

5. What two chemicals represent a safety hazard in this experiment?

32 Paper Chromatography: Separation of Amino Acids

Due day of final (handwritten)

INTRODUCTION

Chemists use many techniques to separate mixtures. Methods such as distillation, sublimation, extraction, and crystallization have been used since ancient times. In recent years the separation of very small quantities of mixtures has become quite important. This is due to increased needs to analyze small samples and for the study of biological systems in which some components appear in small quantities.

Chromatography is a very sensitive separation technique. In chromatography, the mixture to be separated is dissolved in a solvent. The resulting solution, called the *mobile phase*, is then passed over or through another material, called the *stationary phase*. The mixture separates because its components are attracted in different degrees to the two phases. For example, if a component is strongly held by the stationary phase but weakly held by the mobile phase, it will be left behind as the solution passes through the stationary phase. However, a component more strongly attracted to the mobile phase will move along with it as it passes through the stationary phase.

Chromatography is frequently classified by the type of stationary phase used. Adsorption chromatography employs a finely ground solid that adsorbs components of the mixture on its surface. Partition chromatography uses a liquid stationary phase, held in place by an inert solid. The partition (or distribution) of the solute between the solvents gives the technique its name. Ion-exchange chromatography employs a solid ion-exchange resin that releases ions originally within its structure and retains those of the mixture.

Chromatography is also classified according to the mobile phase, which may be either a liquid or a gas—and either a pure substance or a mixture of substances. Gas chromatography is used when the mobile phase is a gas, whether the stationary phase is a solid or a liquid. When the mobile phase is a liquid, however, the technique takes its name from the stationary phase—adsorption, partition, or ion-exchange.

Chromatographic separations that use a liquid mobile phase can also be classified by the type of apparatus used. In column chromatography the solution flows by gravity through a column of circular cross section. In thin-layer chromatography the solution is passed through a thin layer of solid coated on an inert support. Paper chromatography uses paper as the stationary phase. In both thin-layer and paper chromatography the solution moves by capillary action. In high-pressure liquid chromatography the solution is forced through a column by a pump.

Paper Chromatography

Paper is composed essentially of cellulose in which the molecules contain many polar groups. Thus paper has a higher attraction for polar compounds than for nonpolar ones. Components of the mixture that are more soluble in the mobile phase will move further than those that are less soluble. The amount of movement is given by the R_f value, where

Standard (handwritten)

$$R_f = \frac{\text{Distance of component from the starting point}}{\text{Distance of solvent from the starting point}}$$

Tables of R_f values for various solvents are available in the literature. Unfortunately, R_f values depend on a wide variety of variables, including the type of paper (its water content and the direction of its fibers), temperature, distance of the starting point from the solution front, and the manner of production of the chromatogram (ascending, descending, or radial movement). It is difficult to control these variables precisely. Thus it is preferable to run the chromatogram with known components on the same paper as the unknown mixture.

Purpose

The influence of different variables on R_f values will be investigated in this experiment. The technique of paper chromatography will then be used to separate a mixture of amino acids. Paper chromatography is commonly used to separate amino acids because it is relatively quick, easy, and inexpensive. In addition, acceptable results are obtained even by those who have little experience with the technique. The structures of the five amino acids used in this experiment are drawn in Figure 32-1. Even though these amino acids have relatively small differences in their physical properties, they are readily separated by paper chromatography.

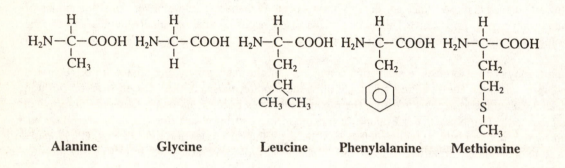

Alanine   **Glycine**   **Leucine**   **Phenylalanine**   **Methionine**

Figure 32-1. Structural Formulas of the Five Amino Acids Used in this Experiment

SAFETY PRECAUTIONS

Review the safety rules on pages 1 and 2.

Some of the solvents used in this experiment are harmful if their vapors are inhaled. Work in a well-ventilated area and wash your hands thoroughly.

Some inks and halide ions are poisonous. Wash your hands thoroughly with soap and running water.

Do not look directly at the ultraviolet light. Do not use the ultraviolet light if you have very sensitive skin.

Avoid inhalation, ingestion, and skin contact with pyridine. It can cause depression of the central nervous system as well as irritation of the skin and respiratory system. Work with extreme caution when using the pyridine-containing solvent system for halide ion separation.

PROCEDURE

A. **Variation of R_f Values**

Separation of Components in Ink

1. Obtain a piece of 7.5 cm filter paper and cut a strip (along the dotted lines) as shown in Figure 32-2. The cuts extend to the center of the paper. This strip will act as a wick in this experiment.

2. Use a capillary tube to place a tiny drop of ink in the center of the paper at the center point marked X. Allow the drop to dry and repeat with three more drops, drying completely after each one.

3. Put water in a 150 mL beaker to within 1 cm of the top. Carefully dry the rim of the beaker. Place the beaker on a clean area of the bench top and put a few drops of water on the bench top beside the beaker.

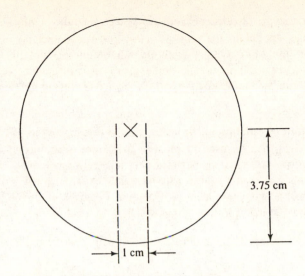

Figure 32-2. Cutting Diagram for Filter Paper

4. Fold the strip of the filter paper at right angles to the rest of the paper. Place the filter paper on the 150 mL beaker so that only the wick dips into the water. The wick should not touch the walls of the beaker.

5. Cover the entire setup with an inverted 800 mL beaker. Allow the chromatogram to develop until the various components of the ink become separated but before the solvent front reaches the edge of the paper.

6. Remove the paper, mark the edge of the solvent front with a pencil, and allow it to dry.

7. Measure the maximum and minimum distance in mm from the starting point for each component of the ink and for the solvent front. Be as accurate as possible.

Effect of Solvent

1. Obtain three large test tubes with corks and support the test tubes vertically.

2. Cut three 15 × 3 cm strips of filter paper and fold them vertically. Make sure the pieces are cut in the same direction from the same piece of filter paper.

3. Spot ink as before (four times with drying between tiny drops) on the crease of each strip 4 cm from one end.

4. Label the strips *a*, *b*, and *c* in pencil on the end furthest from the ink spot.

5. Use an eyedropper to place the solvent indicated in each test tube to a height of 3 cm. Be very careful not to get any solvent on the sides of the test tubes.

Test Tube	Solvent
a	pure water
b	pure acetone
c	acetone + 10% water

6.   Place the appropriately marked filter paper strip in each test tube. Loosely stopper each test tube.

7.   Allow the chromatogram to develop until the solvent is within 2 cm of the top of the paper.

8.   Remove the paper, mark the solvent front with a pencil, and allow it to dry.

9.   Measure the distance in mm from the starting point for each component of the ink and for the solvent front. Be as accurate as you can.

Detection with Ultraviolet Light

1.   If an ultraviolet light source is available, observe each chromatogram under ultraviolet light. *Do not look directly at the ultraviolet light.*

2.   Mark any new spots observed under ultraviolet light with a pencil and measure the distances as before.

DISPOSAL

Solutions: All may be safely flushed down the sink with running water.

B. Halide and Pseudohalide Separation

1.   Be careful during this procedure to avoid touching the filter paper with your hands. Otherwise, halide ions from your hands will be transferred to the filter paper and invalidate the experiment. Use paper towels above and below the paper.

2.   Cut a piece of filter paper approximately 10×20 cm to use in an 800 mL beaker. If another size beaker is used, the width of the paper should be slightly less than the height of the beaker and the length short enough so that the cylinder formed will fit in the beaker without touching the walls (see Figure 32-4).

3.   With pencil draw a line 1 cm from the long edge of the paper and mark the line for F^-, Cl^-, Br^-, I^-, SCN^-, CrO_4^{2-}, and the unknown as shown in Figure 32-3.

4.   Spot the paper as before (four times with drying between tiny drops) making sure the spots are no greater than 5 mm in diameter. Be sure to use the capillary tube in each container of solution and do not mix up the capillaries. Wash your hands with soap and water immediately after you finish spotting the paper.

5.   Allow the filter paper to dry and then separate it into a cylinder with three staples so that the edges of the paper do not touch each other as shown in Figure 32-4.

6.   Fill a beaker with a butanol-aqueous ammonia-pyridine mixture to a depth of 0.5 cm, and place the cylinder in the beaker so that it does not touch the sides. Work in the hood. Pyridine is hazardous.

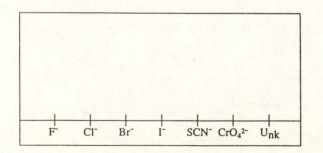

Figure 32-3. Spotting of Halides and Pseudohalides

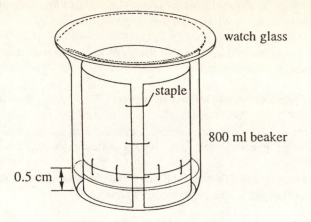

Figure 32-4. Chromatogram Setup

7. Cover the beaker with a watch glass and allow the chromatogram to develop until the solvent front nearly reaches the top of the paper (1 hr). Spot the amino acids (Part C, steps 1 through 5) during this time.

8. Remove the paper, mark the solvent front with a pencil and allow it to dry.

9. Remove the staples and spray the paper with 1.00 M $AgNO_3$ solution so that it is thoroughly wet but not dripping. If you spill $AgNO_3$(aq) solution on your skin, it will cause dark stains that will only wear off in 2 to 3 weeks. Try to be neat so that you are not marked as a careless experimenter.

10. Mark the spots that show, and expose the paper to sunlight as it dries. The silver halide spots will darken on exposure to light, although the silver fluoride spot may take several hours to darken. Place the chromatogram between paper towels in your notebook.

11. Measure the distances in mm from the origin to the center of each spot and to the solvent front.

DISPOSAL

Butano-ammonia-pyridine mixture: Dispose in a waste bottle labeled for this mixture.

C. Amino Acid Separation

1. Be careful to avoid touching the filter paper with your hands. Amino acids on your hands will contaminate the chromatogram.

2. Cut a piece of filter paper as described in Part B, step 2.

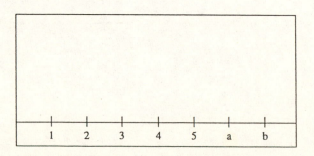

Figure 32-5. Spotting of Amino Acids

3. Mark the paper with pencil as shown in Figure 32-5 for five known amino acids (1 through 5) and two unknown mixtures (*a, b*) on a line 1 cm from the long edge.

4. Spot the paper as before, making sure that the spots are no greater than 5 mm in diameter and that you do not mix up the capillary tubes.

5. Allow the filter paper to dry and staple it as before.

6. In the hood place solvent—a butanol, acetic acid, and water mixture—to a depth of 0.5 cm in a 800 mL beaker and place the paper cylinder in the beaker so that it does not touch the sides (see Figure 32-4).

7. Cover the beaker with a watch glass and allow the chromatogram to develop until the solvent front is three-quarters of the way up the paper (about 90 min).

8. Remove the paper, mark the solvent front with a pencil, and allow it to dry.

9. Remove the staples, spray the paper with ninhydrin solution (work in the hood), and then place the paper in the oven at 80°C for about 3 min. Mark the spots that appear.

10. Measure the distances in millimeters from the origin to the center of each spot and to the solvent front. Also record the color of each spot. (You may have to hold the paper up to the light for help in seeing the colors.) Identify the amino acids in the unknown mixtures.

DISPOSAL

Butanol-acetic acid solution: Dispose in a waste bottle labeled for this solution.

Introduction to Qualitative Analysis
(Experiments 33–38)

Qualitative analysis is that branch of analytical chemistry concerned with the identification of particular substances in a given sample of material. The analysis of inorganic substances involves the study of nonmetallic constituents as anions and metallic constituents as cations. It should be recognized at the outset that the methods we shall use are not those generally used in industrial analytical laboratories. Those methods involve the shortest and most direct procedures, many of which are based on the determination of purely physical properties, such as the observation of spectral lines emitted on strong heating of the material.

The chemical methods of qualitative analysis are used to enable you to apply principles of ionic equilibrium. The solubility product constant, for example, is used to determine the possibility of selectively precipitating certain groups of ions. Acid-base equilibria (buffers) are essential in the control of the concentration of certain precipitating anions. Therefore, the analysis of unknowns provides the motivating factor to make the study of the application of ionic equilibria more effective and interesting.

THE BASIS FOR THE ORGANIZATION OF THE ANALYTICAL SCHEME

Of the approximately 80 metallic elements, 18 of the more common cations are included in the present scheme of analysis. Various approaches to the subdivision of the 18 cations into subgroups have been devised. The most commonly used scheme employs hydrogen sulfide, obtained from thioacetamide, to supply sulfide ion, S^{2-}, and to precipitate many of the metals as sulfides. Consider the approximate solubility product constants of the sulfides of the following metal ions:

Ag_2S	6×10^{-51}	CuS	6×10^{-37}	FeS	6×10^{-19}
Hg_2S	1.0×10^{-47}	SnS	1.0×10^{-26}	ZnS	2×10^{-25}
PbS	3×10^{-28}	HgS	2×10^{-53}	MnS	3×10^{-14}
		Bi_2S_3	1.0×10^{-97}	NiS	3.2×10^{-19}

If the sulfide ion concentration, $[S^{2-}]$, is kept as low as 10^{-22} M, it can be seen that only the sulfides of the first two columns will precipitate, since the ion product $[M^{2+}][S^{2-}]$ must be greater than the K_{sp} of the metal sulfide for precipitation to occur. For example, if the lead ion concentration, $[Pb^{2+}]$, is 0.1 M, then

$$[Pb^{2+}][S^{2-}] = (10^{-1})(10^{-22}) = 10^{-23}$$

which is greater than K_{sp} for PbS, 3×10^{-28}. Therefore, by controlling the sulfide ion concentration, we can selectively precipitate one group of cations from a solution. The $[S^{2-}]$ is controlled by regulating the hydronium ion concentration involved in the H_2S equilibria.

$$H_2S(aq) + H_2O \leftrightharpoons H_3O^+ + HS^- \qquad K_{a1} = \frac{[H_3O^+][HS^-]}{[H_2S]} \qquad = 1.1 \times 10^{-7}$$

$$HS^- + H_2O \leftrightharpoons H_3O^+ + S^{2-} \qquad K_{a2} = \frac{[H_3O^+][S^{2-}]}{[HS]} \qquad = 1.0 \times 10^{-19}$$

$$H_2S(aq) + 2\ H_2O \leftrightharpoons 2\ H_3O^+ + S^{2-} \quad K_{a1}K_{a2} = \frac{[H_3O^+]^2[S^{2-}]}{[H_2S]} \qquad = 1.1 \times 10^{-26}$$

Thus for a solution with constant $[H_2S]$, such as an aqueous solution saturated with hydrogen sulfide, the sulfide ion concentration increases as the hydronium ion concentration decreases (or as the pH increases; see the introduction to Experiment 35).

A further simplification of the problem of separation occurs when we observe that the Ag^+, Hg_2^{2+}, and Pb^{2+} ions can be removed before the sulfide separation, since the chlorides of these cations are only slightly soluble, as shown by their solubility products.

AgCl $\quad 1.8 \times 10^{-10}$ $\qquad\qquad$ Hg_2Cl_2 $\quad 1.3 \times 10^{-18}$ $\qquad\qquad$ $PbCl_2$ $\quad 1.6 \times 10^{-5}$

Ions that are not precipitated as chlorides or sulfides—Ca^{2+}, Ba^{2+}, and Mg^{2+}—can be separated by utilizing the differences in the solubility of their carbonates, and by the control of the carbonate ion concentration. The solubility products of some alkaline earth carbonates follow.

$BaCO_3$ $\quad 5.1 \times 10^{-9}$ $\qquad\qquad$ $CaCO_3$ $\quad 2.8 \times 10^{-9}$ $\qquad\qquad$ $MgCO_3$ $\quad 3.5 \times 10^{-8}$

Study the table below to see how the grouping of metal ions is organized in accordance with the above discussion. The ions of Group I—Na^+ and NH_4^+—are tested for separately and in the presence of other ions that do not interfere. The tests are quite specific for these ions and do not involve precipitation methods.

The Separation of the Metal Ions into Groups

Nitric Acid Solution of the Sample Reagent: HCl or NH_4Cl				
Precipitate AgCl Hg_2Cl_2 $PbCl_2$ (Partly)	Solution (ions of groups III-V) Reagent: H_2S in 0.3 M H^+			
	Precipitate CuS PbS Bi_2S_3 HgS SnS_2	Solutions (ions in groups IV-V) Reagent: $(NH_4)_2S$ in NH_3(aq) solution		
		Precipitate $Al(OH)_3$ FeS, $Fe(OH)_3$ ZnS NiS $Cr(OH)_3$ MnS	Solution (ions of group V) Reagent: $(NH_4)_2CO_3 + NH_4Cl$	
			Precipitate $BaCO_3$ $CaCO_3$	Solution (Mg^{2+}) Reagent: NH_3(aq) + NaH_2PO_4
				Precipitate $MgNH_4PO_4\cdot6H_2O$
(Group II)	(Group III)	(Group IV)	(Group V)	

UNKNOWN ANALYSIS

Each student is required to analyze two or three qualitative analysis unknowns during a seven- to eight-week period. Some details of these unknowns follow.

The first unknown includes Cr^{3+} and a combination of two or more of the ions from the following groups:

I. Soluble group: Na^+, NH_4^+

II. Chloride group: Ag^+, Pb^{2+}, Hg_2^{2+}

III. Hydrogen sulfide group: Pb^{2+}, Cu^{2+}, Sn^{4+}, Bi^{3+}, Hg^{2+}

Cr^{3+} ion might be added to this solution in order to mask the color of Cu^{2+} ion and to allow the student to experience the Group III/IV separation.
The second unknown may include Cu^{2+} and at least two ions from groups IV and V.

IV. Ammonium sulfide group: Fe^{2+}, Fe^{3+}, Al^{3+}, Cr^{3+}, Mn^{2+}, Zn^{2+} and Ni^{2+}

V. Carbonate group: Ba^{2+}, Ca^{2+} and Mg^{2+}

The presence of Cu^{2+} ion provides practice in the separation of Groups III and IV and Cu^{2+} ion also masks the colors of some of the Group IV ions.

Alternatively, one general cation unknown encompassing all five groups may be assigned.

The third unknown involves the analysis of some common negative ions as well as positive ions. In the form of a water-soluble solid salt or salt mixture, the third unknown contains at least one cation from Groups I, IV, or V (as listed above) and one or more anions from the following list.

VI. NO_3^-, NO_2^-, Cl^-, Br^-, I^-, SO_3^{2-}, SO_4^{2-}, PO_4^{3-}, CO_3^{2-}

In preparing for the unknown analysis or to check the unknown tests, you should carry out tests on known solutions available in the laboratory. Known analyses may or may not be required.

None of the unknowns contain more than eight ions. Approximately two to three weeks should be allotted for the analysis of each unknown. After each unknown is completed, a report on available forms is to be completed and turned in. Analysis of the next unknown may begin immediately thereafter. Each unknown will be graded on accuracy with a deduction made for each error (of omission or commission).

LABORATORY TECHNIQUES FOR QUALITATIVE ANALYSIS

The often-repeated admonition—*keep your desk neat and orderly*—will pay big dividends in the saving of your time and in more accurate analyses. Study Figure A. Keep your working space on the bench top and in front of the sink clear of unnecessary equipment. Arrange conveniently the most frequently used items, such as clean test tubes, your wash bottle, stirring rods, medicine dropper, and indicator test paper laid out on a clean towel. Keep dirty test tubes and other articles in one place. Clean and rinse these at the first opportunity so that a stock of clean equipment is always ready. Label any solutions that are to be kept for some time. Keep your laboratory records up to date and do your thinking as you work. To rush through experimental work, with the idea of understanding it later, is an ill-advised attempt at economy of time.

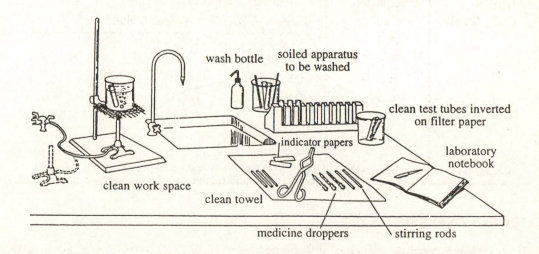

Figure A. Keeping your laboratory worktable neat and in order with the "tools of your trade" conveniently arranged

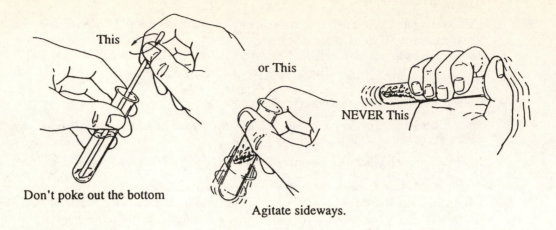

This   or This   NEVER This

Don't poke out the bottom

Agitate sideways.

Figure B. Learning to use good technique in mixing solutions

The Volume of Solutions

Use only the small volumes specified. You can estimate most volumes with sufficient precision by counting drops (15 to 20 drops per mL) or by estimating the height of the solution in the 10 cm test tube (capacity 8–10 mL). Use the 10 mL graduated cylinder only when more accurate volumes are required.

The Handling of Solutions

Learn to be clean and efficient. When mixing solutions, use a stirring rod (Figure B); do not invert the test tube with your dirty thumb as the stopper. When heating solutions, avoid loss by bumping (Figure C). Do not try to boil more than 2 mL of solution in a 10 cm test tube; transfer large quantities to a 15 cm test tube, a small beaker, or a casserole. Solutions in test tubes may be heated safely by immersion in a beaker of boiling water.

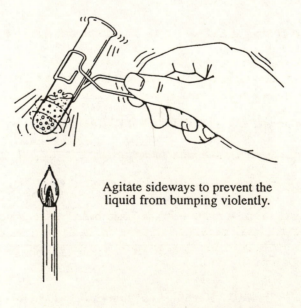

Agitate sideways to prevent the
liquid from bumping violently.

**Figure C. Improving your technique when
you evaporate a solution in a test tube**

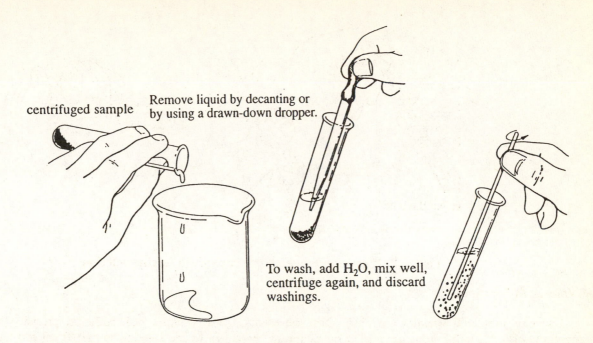

centrifuged sample

Remove liquid by decanting or by using a drawn-down dropper.

To wash, add H$_2$O, mix well, centrifuge again, and discard washings.

Figure D. Separating and washing a precipitate that has been centrifuged

The Separation of Precipitates

The centrifuge is used to separate precipitates, rather than simply allowing the solid to settle by gravity. Centrifugation is faster (1–2 min centrifuging time usually is sufficient) and has been specified throughout these procedures. To wash a centrifuged precipitate, pour off the solution, let it drain, and touch off the last drop against the side of the receiver. Even if it is to be discarded, a liquid should be decanted into another container and not into the sink. If this technique is followed, any precipitate that inadvertently is decanted can be recovered. Mix the precipitate with 2 to 3 mL of distilled water, again centrifuge this, and pour off the liquid. This may be repeated (Figure D).

The Preparation of Hydrogen Sulfide by the Thioacetamide Method

Thioacetamide, an organic reagent used in volumes of about 10-12 drops of 1 M CH$_3$CSNH$_2$, is added directly to the 2–4 mL sample to be saturated with H$_2$S. The test tube containing the sample is then immersed in boiling water in a beaker for 10 min (Figure E). The resulting hydrolysis reaction, in acid solution, is

$$CH_3CSNH_2(aq) + HOH \longrightarrow CH_3CONH_2(aq) + H_2S(aq) \qquad \text{[Qual.1]}$$

Further hydrolysis, especially in basic solution, occurs.

$$CH_3CONH_2(aq) + HOH \longrightarrow NH_4^+ + CH_3COO^- \qquad \text{[Qual.2]}$$

The ammonium acetate produced in the basic hydrolysis, equation [Qual.2], exhibits a buffering effect which can result in pH adjustment being more difficult. Also, the precipitation of metallic sulfides (MS) always produces H$^+$ in the solution, in proportion to the amount of sulfide precipitate formed according to equation [Qual.3]:

$$M^{2+} + H_2S(aq) \longrightarrow MS(s) + 2 H^+ \qquad \text{[Qual.3]}$$

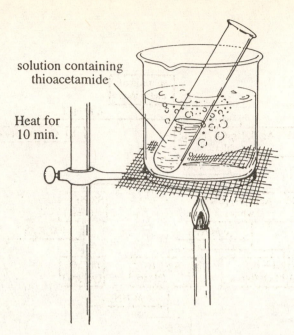

solution containing
thioacetamide

Heat for
10 min.

Figure E. Generating a saturated H₂S solution with the thioacetamide method

Although other methods can be used to generate H_2S, the thioacetamide method is preferred by many chemists, not only because of the freedom from odors in the laboratory but also because the slow rate of production of H_2S by hydrolysis favors the formation of more crystalline precipitates, which are more easily filtered or centrifuged. The rate of hydrolysis is very slow in neutral solution, so the method is not satisfactory in specific experiments where this is an essential condition of the precipitation. The sulfide separations in the regular procedure, however, are carried out in either a definitely acid or a definitely basic solution.

FLOW CHARTS

Beginning with Group II, a flowchart should be completed before beginning work on each group.   A sample chart for Group II is shown in Figure F.

You should also be sure that you understand the chemical reactions that occur during each test. Both a general knowledge of the qualitative analysis scheme and an understanding of all reactions involved are considered to be an essential part of the course.

REPORT FORMS

Following are sample report forms for (a) a qualitative analysis unknown consisting of selected ions (NH_4^+, Pb^{2+}, and Ag^+) from Groups I and II and (b) an unknown solid salt. Report form blanks similar to these are available in the appendix of this manual and should be used for each unknown. The format of the report forms is also recommended as the original record in your laboratory notebook.

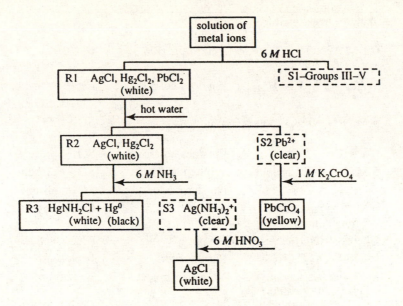

Figure F. Flowchart for Group II

Sample Report A   Name _____ Section _____

Substance	Reagent	Result	Conclusion
Unknown	Sodium flame test	No orange flame	Na^+ absent
Unknown	Ammonium test	Litmus turns blue	NH_4^+ present
Unknown	6 M HCl	White ppt (R1)	Pb^{2+}, Hg_2^{2+}
		Clear solution (S1)	Ag^+ possible
R1 hot water	Clear solution (S2)	Pb^{2+} possible + white ppt (R2)	
S2	1 M K_2CrO_4	Yellow ppt	Pb^{2+} present
R2 6 M NH_3	Clear solution (S3)	Hg_2^{2+} absent	
S3	6 M HNO_3	White ppt	Ag^+ present

Ions present: NH_4^+, Pb^{2+}, Ag^+

Sample Report B   Name _____ Section _____

Substance	Reagent	Result	Conclusion
Unknown is a light green solid	Water + heat	Pale green solution (Sa)	
Sa	0.1 M $AgNO_3$	Yellow ppt (Ra)	Cl^-, Br^-, I^-, CO_3^{2-}, SO_3^{2-}, PO_4^{3-} possible
Ra	Water + 3 M HNO_3	Ppt does not dissolve	Cl^-, Br^-, I^- possible CO_3^{2-}, SO_3^{2-}, PO_4^{3-} absent (?)
Sa	3 M NH_3 + $BaCl_2$ soln	White ppt	SO_4^{2-}, PO_4^{3-}, SO_3^{2-}, CO_3^{2-} possible
Rb	6 M HCl	Ppt does not dissolve	SO_4^{2-} present
Sa	6 M HCl + $BaCl_2$ soln	White ppt No sharp odor	SO_4^{2-} present SO_3^{2-} absent
Sa	6 M HCl; pass gas through limewater	No ppt	CO_3^{2-} absent
Sa	Cl^- test (HNO_3 & $AgNO_3$)	No ppt	Cl^- absent
Sa	0.1 M $FeCl_3$ $C_2H_3Cl_3$	Purple color	I^- present
Sb	Separate; use aqueous layer, add Cl_2 water	No brown color	Br^- absent
Sa	Make acidic w/HNO_3 + $(NH_4)_2MoO_4$	No yellow ppt	PO_4^{3-} absent
Sa	Cold conc. H_2SO_4	No brown ring	NO_2^- absent
Sa	AgAc soln + HCl	Yellow ppt and solution (Sa)	Br^- or I^- present
Sc	$FeSO_4$ + H_2SO_4	No brown ring	NO_3^- absent
Sa	6 M NaOH + wet red litmus	Litmus turns blue	NH_4^+ present
Sa	Flame test	Green flame	Na^+ absent
Sa	3 M NH_4Cl + NH_3(aq)	Solution (Sa) but no ppt	Fe^{3+}, Al^{3+}, Cr^{3+} probably absent
Sd	Sat'd with H_2S	Clear soln (S9) Black ppt (R9)	Group IV ion present
R9	1 M Na_2SO_4 + 1 M $NaHSO_4$	Clear soln (S10) Black ppt (R10)	Zn^{2+} or Ni^{2+} present
R10	H_2O + 6 M HCl	Clear soln (S11) + black ppt (R10)	Ni^{2+} present
S11	0.1 M NaHS + 3 M NH_4Ac	No ppt	Zn^{2+} absent
R11	HNO_3, dry + H_2O + NH_3(aq) + DMG	Red ppt	Ni^{2+} confirmed
S10	Boil + H_2O_2 + NaOH	Soln (S12) no ppt	Mn^{2+}, Fe^{2+}, Fe^{3+} absent
S12	HCl + NH_3(aq) to make basic	Soln (S13) No ppt	Al^{3+} absent
S13	HAc + 0.1 M $Pb(Ac)_2$	No ppt	Cr^{3+} absent
S9	3 M $(NH_4)_2CO_3$	Soln (S14) no ppt	Ba^{2+}, Ca^{2+} absent
S14	6 M NH_3(aq)	White ppt (Rc)	Mg^{2+} present
Rc	dissolved in HCl + magnesium reagent	Blue lake	Mg^{2+} confirmed

Ions present: SO_4^{2-}, I^-, NH_4^+, Ni^{2+}, Mg^{2+}

Prelab   Name _____ Section _____

Introduction to Qualitative Analysis

1. How is the concentration of S^{2-} ion controlled in the reaction mixture?

2. If the sulfide ion concentration is slowly increased in a solution containing Cu^{2+} and Hg^{2+}, which ion will begin to precipitate first?

3. List four anions that are used in this qualitative analysis scheme to precipitate metal cations.

4. Which carbonate in Group V is most insoluble?

5. Are bismuth and tin ions precipitated from the unknown solution as chlorides, sulfides, hydroxides, carbonates, or none of these?

6. What can be done to prevent bumping when heating a solution in a test tube over a Bunsen burner?

7. What is centrifugation? What is its purpose in the qualitative analysis experiments?

8. How is thioacetamide able to act as a source of H_2S?

9. What are the advantages of using thioacetamide as a source of H_2S?

10. Using the flow chart in Figure F, describe which ion(s) remain(s) as a solid chloride precipitate and which dissolve(s) into solution when hot water is added to a sample of $AgCl$, Hg_2Cl_2, and $PbCl_2$.

33 Group I: The Soluble Group

INTRODUCTION

The two ions included in Group I are sodium ion, Na^+, and ammonium ion NH_4^+. Sodium ion is representative of the ions of the very active alkali metals located in Group IA of the periodic table. The salts of these metals are nearly all readily soluble in water. Although ammonium ion is not an alkali metal ion, its salts likewise are nearly all soluble. In fact, NH_4^+ is often compared to K^+ in that many of its properties are quite similar to those of K^+.

The identification tests for NH_4^+ and Na^+ must be carried out on separate portions of an unknown solution. These ions are unique in this respect. No separation tests are necessary. Their tests are simple and convenient.

PROCEDURE FOR THE ANALYSIS OF AN UNKNOWN SOLUTION FOR NH_4^+ AND Na^+

Note: The "Introduction to Qualitative Analysis" (pp. 281–288) should be studied thoroughly before starting any of the following tests.

Test for Ammonium Ion

$$NH_4^+(aq) + OH^-(aq) \leftrightarrows NH_3(g) + H_2O(l) \qquad [33.1]$$

Place 3 mL of the solution to be tested in an evaporating dish, and add 1 to 2 mL of 6 M NaOH, equation [33.1]. Quickly cover the dish with a watch glass, on the underside of which is attached a moist strip of red litmus paper. Warm the solution very gently (over hot water, Figure 33-1, not with a burner flame or a hot plate) to liberate any ammonia present as the gas. Avoid boiling, which would contaminate the litmus with spray droplets of the NaOH solution. An even, unspotted blue color on the litmus (reaction [33.1]) confirms the presence of ammonium ion. The characteristic odor of ammonia, observed soon after the solution is warmed, also serves as a positive test. (Be very cautious in bringing your nostrils close to a hot sodium hydroxide solution, however, as it might be superheated, and spatter in your face or eyes.)

Test for Sodium Ion

$$Na^+(aq) \rightarrow Na^+(g) \qquad [33.2]$$

Clean a wire of nickel-chromium alloy as shown in Figure 33-2 until no color is observed when the wire is held in the flame of the Bunsen burner. Dip the clean wire in the solution to be tested, then hold it in the nonluminous flame of the burner. A luminous fluffy yellow-orange coloration in the flame that persists for 30 sec or more indicates the presence of Na^+ (rxn. 33.2) Care must be taken since this test is extremely sensitive. Traces of sodium ion get into solutions as a result of contacting glassware; similarly contact of the wire with one's fingers will cause some salt to be transferred to the wire; further, nearly all reagents contain a trace of sodium ion. The result is that most solutions give what might be considered to be a weakly positive test for Na^+. A comparison with a solution known to contain sodium ion as well as with a solution containing trace amounts of Na^+ is very helpful in making the decision whether sodium ion is present in the unknown solution.

DISPOSAL

Aqueous solutions: Dispose in a common waste bottle labeled for *qualitative analysis—solutions.*

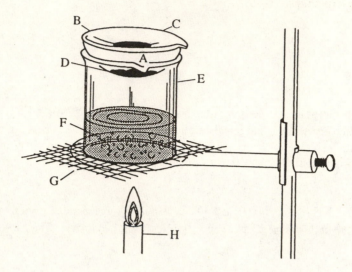

Figure 33-1. The Test for Ammonium Ion:
A **is red litmus paper moistened and placed on the underside of the watch glass,** *B.*
C **is an evaporating dish containing the sample and some sodium hydroxide solution,** *D.*
E **is a beaker,** *F* **is hot water,** *G* **is a wire screen, and** *H* **is a burner.**

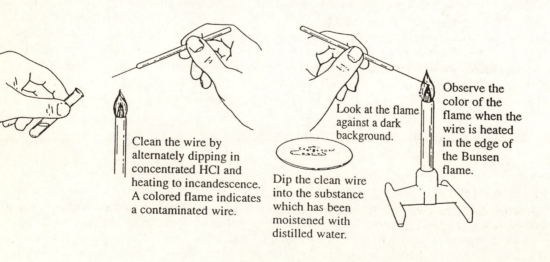

Clean the wire by alternately dipping in concentrated HCl and heating to incandescence. A colored flame indicates a contaminated wire.

Dip the clean wire into the substance which has been moistened with distilled water.

Look at the flame against a dark background.

Observe the color of the flame when the wire is heated in the edge of the Bunsen flame.

Figure 33-2. The Flame Test for Sodium Ion:
(a) Cleaning the Wire and (b) Observing the Flame Colorations

Prelab   Name _____ Section _____

Group I: The Soluble Group

1. Why are Na^+ and NH_4^+ ions not separated by precipitation with the addition of an anion, like the other metal ions in Experiments 34–37?

2. What reacts with the litmus paper to turn it blue in the first test?

3. Why is it necessary to heat the ammonium test solution gently without boiling?

4. How should the flame test wire be cleaned before testing for Na^+?

5. When doing the flame test for sodium, what should you do if you are not certain if your results indicate the presence of sodium ions?

34 Group II: The Chloride Group

INTRODUCTION

The precipitating reagent for the chloride group is chloride ion in an acidic solution. Only three of the 18 metal ions form a precipitate with this reagent: silver ion (Ag^+), mercury(I) ion (Hg_2^{2+}), and lead ion (Pb^{2+}). Since lead chloride is somewhat soluble, it is not completely precipitated here, and a low concentration of lead ion appears also in Group III.

The behavior of mercury salts is unusual and needs some explanation. Mercury exhibits three oxidation states: zero in free mercury, $Hg°$, +1 in mercury(I) ion, Hg_2^{2+}, and +2 in mercury(II) ion, Hg^{2+}. In some situations, the intermediate mercury(I) ion, Hg_2^{2+}, is unstable, some being reduced to the metal, and part oxidized to Hg^{2+}.

$$Hg_2^{2+} \longrightarrow Hg° + Hg^{2+} \qquad [34.1]$$

When aqueous ammonia is added to mercury(II) chloride solution, a white "ammonolysis" product, $HgNH_2Cl$, is formed. This is analogous to the partial hydrolysis of $HgCl_2$ to form a basic salt. Compare the two equations.

$$HgCl_2(aq) + HOH \longrightarrow Hg(OH)Cl(s) + HCl \qquad [34.2]$$

$$HgCl_2(aq) + HNH_2 \longrightarrow Hg(NH_2)Cl(s) + HCl \qquad [34.3]$$

If mercury(I) chloride is treated with ammonia, part of it is oxidized to the white mercury(II) amidochloride (equation [34.3]), and part is reduced to black, finely divided mercury (equation [34.1]). The mixed precipitates appear black or gray.

PROCEDURE FOR THE ANALYSIS OF AN UNKNOWN SOLUTION FOR Ag^+, Pb^{2+}, AND Hg_2^{2+}

Before starting analysis, complete the Chloride group flowchart (Figure 34-0).

Note: You may experience some difficulty in identifying the ions in your unknown, or you may be uncertain about whether a quantity of precipitate indicates that an ion is present or not. A side-by-side comparison of tests with a solution known to contain the suspected ion will be very helpful in such cases.

Precipitation of the Group

$$Ag^+(aq) + Cl^-(aq) \longrightarrow AgCl \ (s, \textit{white}) \qquad [34.4]$$

$$Pb^{2+}(aq) + 2 \, Cl^-(aq) \longrightarrow PbCl_2 \ (s, \textit{white}) \qquad [34.5]$$

$$Hg_2^{2+}(aq) + 2 \, Cl^-(aq) \longrightarrow Hg_2Cl_2 \ (s, \textit{white}) \qquad [34.6]$$

Add 3 drops of 6 M HCl to 3 mL of the test solution, and mix thoroughly by stirring the mixture with a glass stirring rod. Centrifuge, decant the centrifugate (solution), and test for completeness of precipitation by adding 1 to 2 more drops of 6 M HCl. If no more precipitate forms, precipitation is complete and the liquid (S1) may be discarded if only Group II ions are possible in this unknown. If the unknown possibly contains ions from groups III–V, S1 must be saved. In this case, stopper and label the test tube containing S1 for later analysis. Wash the residue (R1) with 1 to 3 mL of distilled water from your wash bottle, stir, centrifuge, decant, and discard the wash liquid. The residue R1 is now ready for further analysis of the ions in Group II. If no residue forms upon addition of 6 M HCl, none of the three ions in Group II is present and further analysis of Group II is unnecessary.

Test for Lead Ion

$$PbCl_2(s) \xrightarrow{\Delta} Pb^{2+}(aq) + 2\ Cl^-(aq) \qquad [34.7]$$

$$Pb^{2+}(aq) + CrO_4^{2-}(aq) \longrightarrow PbCrO_4\ (s,\ yellow) \qquad [34.8]$$

Add 3 to 4 mL of distilled water to residue R1 and heat the mixture to near boiling while stirring. Centrifuge the mixture and decant the solution (S2). Wash the remaining residue (R2) with 10 mL of boiling distilled water in portions, centrifuge, and discard these washings. The washings from this precipitation must be discarded in a container labeled qualitative analysis—solutions. Add several drops of 1 M K_2CrO_4 to the centrifugate (S2). A yellow precipitate (reaction [34.8]) confirms the presence of Pb^{2+}.

Test for Mercury(I) Ion

$$AgCl(s) + 2\ NH_3(aq) \longrightarrow Ag(NH_3)_2^+(aq) + Cl^-(aq) \qquad [34.9]$$

$$Hg_2Cl_2(s) + 2\ NH_3(aq) \longrightarrow Hg(NH_2)Cl\ (s,\ white) + Hg^\circ\ (l,\ black) + NH_4^+(aq) + Cl^-(aq) \qquad [34.10]$$

Add 1 mL of 6 M $NH_3(aq)$ and then 2 mL of water with stirring to residue R2. Centrifuge and decant the centrifugate (S3) into a small test tube. A gray to black residue (reaction [34.10]) confirms the presence of Hg_2^{2+}. This residue should be discarded in the waste bottle labeled qualitative analysis—solids.

Test for Silver Ion

$$Ag(NH_3)_2^+(aq) + Cl^-(aq) + 2\ H^+(aq) \longrightarrow AgCl\ (s,\ white) + 2\ NH_4^+(aq) \qquad [34.11]$$

If S3 is not perfectly clear, it may be due to colloidal Pb(OH)Cl resisting centrifugation. In this case, recentrifuge the solution as often as necessary until a clear liquid is obtained. Then acidify S3 by adding 6 M HNO_3 dropwise with stirring until a piece of blue litmus paper turns red after being contacted with the stirring rod. A white precipitate (reaction [34.11]) confirms the presence of Ag^+.

DISPOSAL

Aqueous solutions: Dispose in a common waste bottle labeled for *qualitative analysis—solutions.*
Solids: Dispose in a waste bottle labeled *qualitative analysis—solids.*
Wash solutions/solids mixtures: Dispose in a waste bottle labeled *qualitative analysis—wash mixtures.*
All of these wastes will be collected, treated, and stored for volume reduction and proper disposal by qualified laboratory personnel.

Figure 34-0. Flowchart: The Chloride Group

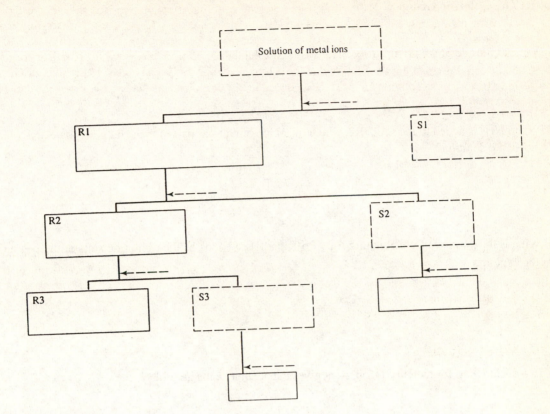

Prelab   Name _____ Section _____

Group II: The Chloride Group

1. Which chloride of the Group II metal ions is least soluble?

2. When NH_3 is added to the solid, R2, which metal ion remains as part of the solid precipitate and which ion goes into solution?

3. What is the purpose of precipitating Ag^+, Pb^{2+}, and Hg_2^{2+} ions as their chloride salts at the beginning of the experiment?

4. In a positive test for mercury(I) ion, what gives the precipitate the gray-black color?

5. If your unknown sample contains metal ions to be analyzed in Experiment 35, which solution(s) must be saved from Experiment 34?

35 Group III: The Hydrogen Sulfide Group

INTRODUCTION

Hydrogen Sulfide—A Weak Diprotic Acid

When hydrogen sulfide is dissolved in water at ordinary atmospheric pressure, it gives a saturated solution which is about 0.1 M H_2S. This behaves as a weak acid, and since there are two replaceable hydrogen atoms we may represent its ionization in steps.

$$H_2S + H_2O \leftrightarrows H_3O^+ + HS^- \tag{35.1}$$

$$HS^- + H_2O \leftrightarrows H_3O^+ + S^{2-} \tag{35.2}$$

The extent of the ionization of hydrogen sulfide ion in [35.2] is much less than that of H_2S according to [35.1], for it involves the removal of a positive hydrogen ion from the negative ion HS^-; the first hydrogen ion has only to be separated from the neutral molecule, H_2S.

$$\frac{[H_3O^+][HS^-]}{[H_2S]} = K_{a1} = 1.1 \times 10^{-7} \tag{35.3}$$

$$\frac{[H_3O^+][S^{2-}]}{[HS^-]} = K_{a2} = 1.0 \times 10^{-19} \tag{35.4}$$

Note that K_2 is about 10^{12} times smaller than K_1, corresponding to the much smaller extent of ionization in [35.2]. The expression $[H_3O^+]$ refers to the same value, of course, in both equations [35.3] and [35.4], since both equilibria are present in the same solution. Most of the H_3O^+ comes from the first ionization. Likewise, the HS^- concentration (from the first ionization) is about equal to the H_3O^+ concentration and is not nearly as small as the S^{2-} concentration, which results solely from the second ionization. If we multiply equation [35.3] by equation [35.4] and cancel out the common factor $[HS^-]$, we obtain

$$\frac{[H_3O^+]^2[S^{2-}]}{[H_2S]} = K_{a1}K_{a2} = K_a = (1.1 \times 10^{-7})(1.0 \times 10^{-19}) = 1.1 \times 10^{-26} \tag{35.5}$$

In a solution saturated with hydrogen sulfide at atmospheric pressure the concentration of molecular H_2S is about 0.1 M. We may, therefore, simplify equation [35.5] as follows:

$$\frac{[H_3O^+]^2[S^{2-}]}{[0.1]} = 1.1 \times 10^{-26}$$

$$[H_3O^+]^2[S^{2-}] = 1.1 \times 10^{-27} \tag{35.6}$$

This is a very important relationship, as it shows us, for instance, that if the hydronium ion concentration in a solution is increased tenfold, the sulfide ion concentration in the solution is thereby reduced a hundredfold. That is, the sulfide ion concentration in a solution saturated with hydrogen sulfide is inversely proportional to the square of the hydronium ion concentration in that solution. *We may thus control the maximum sulfide ion concentration in a solution by controlling the hydronium ion concentration.*

For example, in a 10^{-2} M H_3O^+ solution, we can calculate the maximum sulfide ion concentration resulting when hydrogen sulfide saturates the solution, by substituting in equation [35.6].

$$[10^{-2}]^2[S^{2-}] = 1.1 \times 10^{-27}$$

$$[S^{2-}] = \frac{1.1 \times 10^{-27}}{10^{-4}} = 1.1 \times 10^{-23} \text{ M}$$

Similarly, in a 0.30 M H_3O^+ solution, such as that used when precipitating the hydrogen sulfide group in acid solution, we may calculate the maximum sulfide ion concentration.

$$[0.30]^2[S^{2-}] = 1.1 \times 10^{-27}$$

$$[S^{2-}] = \frac{1.1 \times 10^{-27}}{0.090} = 1.2 \times 10^{-26}$$

Note in these two examples that increasing the hydronium ion concentration 30 times (from 10^{-2} to 0.3 M) has resulted in a decrease of about a thousandfold in the sulfide ion concentration. The effect this large difference in sulfide ion concentration has on the ability of hydrogen sulfide to precipitate a metallic sulfide is discussed in the next section.

The Precipitation of Metallic Sulfides

In a saturated solution of copper(II) sulfide, we have the equilibrium

$$CuS(s) \leftrightarrows Cu^{2+}(aq) + S^{2-}(aq). \tag{35.7}$$

According to the solubility product principle, the product of the concentrations of the ions in the solution cannot be greater than the value of the solubility product constant.

$$[Cu^{2+}][S^{2-}] = K_{sp} = 6.3 \times 10^{-36} \tag{35.8}$$

The amount of copper(II) ion that can remain in solution depends, therefore, on the sulfide ion concentration, and this in turn depends on the hydronium ion concentration. For a 0.3 M H_3O^+ solution, the sulfide ion concentration cannot be greater than 1.2×10^{-26} M. (See the above examples.) From this, we may calculate the maximum copper(II) ion concentration that could remain in such a solution. Substituting in equation [35.8], we get the following relationships:

$$[Cu^{2+}][1.2 \times 10^{-26}] = 6.3 \times 10^{-36}$$

$$[Cu^{2+}] = \frac{6.3 \times 10^{-36}}{1.2 \times 10^{-26}} = 5.3 \times 10^{-10} \text{ M}$$

The precipitation of copper(II) ion from such a solution, therefore, is quantitative (complete).
Compare this with the somewhat more soluble iron(II) sulfide, for which, in a saturated solution,

$$[Fe^{2+}][S^{2-}] = K_{sp} = 6.3 \times 10^{-18} \tag{35.9}$$

Again, considering a 0.3 M H_3O^+ solution, in which $[S^{2-}] = 1.2 \times 10^{-26}$, we get the following relationships.

$$[Fe^{2+}][1.2 \times 10^{-26}] = 6.3 \times 10^{-18}$$

$$[Fe^{2+}] = \frac{6.3 \times 10^{-18}}{1.2 \times 10^{-26}} = 5.3 \times 10^8$$

Thus a very high concentration of iron(II) ion can remain in solution without being precipitated. However, if the solution is made basic with NH_3(aq) and then saturated with hydrogen sulfide – say at a pH of about 8 – the sulfide ion concentration can increase to about 10^{-7} M. The maximum Fe^{2+} that can now remain in solution may be calculated.

$$[Fe^{2+}][10^{-7}] = 6.3 \times 10^{-18} \quad or \quad [Fe^{2+}] = 6.3 \times 10^{-11}$$

Thus iron(II) ion is almost completely precipitated in a slightly basic solution, but not at all in a 0.3 M H^+ solution.

The very insoluble metallic sulfides are precipitated along with copper(II) sulfide in the hydrogen sulfide group, whereas those that are slightly more soluble may be precipitated along with iron(II) sulfide in the ammonium sulfide group. The readily soluble sulfides of the metals in Group V will not be precipitated in either case. This is the basic principle of separation of Groups III, IV, and V in the qualitative analysis scheme.

Note on the Insolubility of Metallic Sulfides

The solubility product constants of many of these quite insoluble sulfides are not known more precisely than to the nearest power of 10, and there is considerable variation in their listed values. Calculations such as the preceding one should be regarded as only approximate. They do serve, however, to show the principles on which the separation of metallic sulfides is based.

Furthermore, the degree of insolubility seems to vary with time after precipitation. Students are advised to avoid keeping precipitates from one laboratory period to the next unless they centrifuge and separate the precipitate from the solution.

The Separation of the Hydrogen Sulfide Group into Subgroups

Amphoteric Sulfides

As we have just seen, the hydrogen sulfide group includes all those sulfides that are so insoluble as to be almost completely precipitated in 0.3 M H_3O^+ solution by hydrogen sulfide. This includes five of the 18 metal ions to be studied. For convenience, these ions may be subdivided into two smaller groups for individual tests. This is accomplished by the treatment of the mixed sulfide precipitate with a high sulfide ion concentration. Using a sodium hydrogen sulfide solution, we find that the sulfides of mercury(II) ion and tin(IV) ion dissolve as the corresponding thio complexes. (If ammonium sulfide, which gives a lower sulfide ion concentration due to hydrolysis, is used, the mercury(II) sulfide remains undissolved with the "copper" subgroup.) The theory of such separations follows.

Just as the hydroxides of certain metals are amphoteric, dissolving in an excess of hydroxide ion to form hydroxo- complex ions, so also the sulfides of some of the metals dissolve in a high concentration of sulfide ion to form sulfide complex ions. The addition of only a slight excess of hydroxide ion, or of sulfide ion, to a tin(IV) salt solution forms the expected precipitates.[*]

---

[*]An ion with such a high positive charge as 4+ has a very strong tendency to coordinate with any negative ions in solution and will seldom, if ever, be present as the simple ion. In a hydrochloric acid solution, tin(IV) salts form the chloride complex ions, $[SnCl_6]^{2-}$, so that the reactions with hydroxide or sulfide ions really constitute replacements.

$$[SnCl_6]^{2-}(aq) + 4\ OH^-(aq) \leftrightarrows Sn(OH)_4(s) + 6\ Cl^-(aq)$$

$$[SnCl_6]^{2-}(aq) + 2\ S^{2-}(aq) \leftrightarrows SnS_2(s) + 6\ Cl^-(aq)$$

$$Sn^{4+}(aq) + 4\ OH^-(aq) \leftrightarrows Sn(OH)_4(s) \qquad\qquad [35.10]$$

$$Sn^{4+}(aq) + 2\ S^{2-}(aq) \leftrightarrows SnS_2(s) \qquad\qquad [35.11]$$

But with a high concentration of these ions, the precipitates redissolve to form complexes.

$$Sn(OH)_4(s) + 2\ OH^-(aq) \leftrightarrows [Sn(OH)_6]^{2-}(aq) \qquad\qquad [35.12]$$

$$SnS_2(s) + S^{2-}(aq) \leftrightarrows [SnS_3]^{2-}(aq) \qquad\qquad [35.13]$$

The addition of hydronium ion to the above strongly basic solutions decreases the concentration of hydroxide ion, or of sulfide ion, and reprecipitates tin(IV) hydroxide or tin(IV) sulfide, respectively. In a high concentration of hydrochloric acid, both precipitates redissolve again, owing not only to the action of the hydronium ion in forming water and hydrogen sulfide, respectively, but also to the high chloride ion concentration, which forms the complex ion $[SnCl_6]^{2-}$.

The situation is a very interesting one: tin(IV) sulfide dissolves readily in a sulfide solution, while tin(II) sulfide does not dissolve. Sulfide ion, a Brønsted base,* coordinates with tin(IV) sulfide because SnS_2 is a sufficiently strong Brønsted acid to react. Tin(II) sulfide is not a strong enough acid to react with sulfide ion. This behavior is in accord with the more acidic character of an element in its higher oxidation state. Likewise, nitrous acid (HNO_2) and sulfurous acid (H_2SO_3) are weak acids, whereas the corresponding acids in which the elements are more highly oxidized, nitric acid and sulfuric acid, are strong acids. It is for this reason that, in the analytical procedure, an oxidizing agent, hydrogen peroxide, is added to oxidize any tin to tin(IV) ion before adding sulfide ion.

The Oxidation of Hydrogen Sulfide and Metallic Sulfides

Thus far we have dealt with hydrogen sulfide only as a weak acid. It contains sulfur in its lowest oxidation state, −2, and is also a good reducing agent, reacting readily with any strong oxidizing agents present. The sulfide ion is nearly always oxidized to free sulfur, together with some sulfate ion that forms by further oxidation.

Nitric acid, or nitrate ion in acidic solution, is nearly always present in the solution to be tested for Group III ions. If the solution is quite dilute and not too hot, these nitrates do not interfere as oxidizing agents. In more concentrated solutions, especially when heated, sulfur always forms when hydrogen sulfide is produced in the solution. The reaction is

$$3\ H_2S(g) + 2\ H^+ + 2\ NO_3^- \longrightarrow 3\ S(s) + 2\ NO(g) + 4\ H_2O \qquad\qquad [35.14]$$

This also explains why nitric acid is a better solvent than hydrochloric acid for most metallic sulfides. To dissolve copper(II) sulfide, we must decrease the concentration of either the copper(II) ion or the sulfide ion, so as to shift the equilibrium for equation [35.7]

$$CuS(s) \leftrightarrows Cu^{2+}(aq) + S^{2-}(aq)$$

to the right. The sulfide ion concentration is decreased more completely by oxidizing it to free sulfur with nitric acid than it is by forming the weak acid hydrogen sulfide with hydrochloric acid. The reaction is

---

*This is in accord with the definition of a Brønsted base as any substance that can accept protons, H^+. Thus sulfide ion, carbonate ion, acetate ion, etc., as well as hydroxide ion, are bases.

$$3\ CuS(s) + 8\ H^+(aq) + 2\ NO_3^-(aq) \longrightarrow 3\ Cu^{2+}(aq) + 3\ S(s) + 2\ NO(g) + 4\ H_2O. \qquad [35.15]$$

Strong hydrochloric acid, however, dissolves some quite insoluble sulfides by both decreasing the sulfide ion concentration to form the weak acid and decreasing the metal ion concentration by the formation of a chloro complex ion (such as $[CuCl_4]^{2-}$, $[SnCl_4]^{2-}$, $[SnCl_6]^{2-}$, and $[FeCl_4]^-$). With tin(IV) sulfide, the equation for the reaction is

$$SnS_2(s) + 4\ H^+(aq) + 6\ Cl^-(aq) \longrightarrow [SnCl_6]^{2-}(aq) + 2\ H_2S(g). \qquad [35.16]$$

Mercury(II) sulfide is so extremely insoluble that it does not dissolve in either nitric acid or hydrochloric acid separately, but dissolves readily in the combination of these acids, known as aqua regia. This reagent has a powerful solvent action because in addition to the oxidizing action of nitric acid there is also a high concentration of chloride ion, thus reducing the concentration of mercury(II) ion by the formation of the complex ion $[HgCl_4{}^{2-}]$. With concentrated acids, we have the reaction

$$HgS(s) + 4\ H^+ + 2\ NO_3^- + 4\ Cl^- \longrightarrow [HgCl_4{}^{2-}] + S(s) + 2\ NO_2(g) + 2\ H_2O. \qquad [35.17]$$

SAFETY PRECAUTIONS

Review the safety rules on pages 1 and 2.

Wear safety goggles at all times.

Be particularly careful when heating and mixing chemicals.

Use the hood to exhaust any fumes that might be released when carrying out the following reactions.

Use extra care when using *30% H_2O_2*. If any gets on the skin, wash immediately with water for at least 5 min.

PROCEDURE FOR THE ANALYSIS OF A SOLUTION FOR Cu^{2+}, Pb^{2+}, Sn^{4+}, Bi^{3+}, and Hg^{2+}

Before starting analysis, complete the hydrogen sulfide group flowchart (Figure 35-0).

Since the test solution (S1 from Group II or one containing Group III ions only) is already strongly acidic, it must be neutralized first with $NH_3(aq)$ and then adjusted to 0.3 M H_3O^+ by the addition of a controlled quantity of acid. Accordingly, neutralize 3 mL of the acidic test solution (S1 from Group II or Group III solution) by adding 6 M $NH_3(aq)$ dropwise with stirring until one drop turns red litmus blue (stirring rod touched to litmus paper). Then add 1 drop of 6 M HCl for each mL of test solution. Test with blue litmus paper. If the solution is not acidic continue to add HCl dropwise with stirring until an acidic litmus test is obtained. (If a precipitate formed during the addition of the NH_3 add HCL dropwise until it completely dissolves.)

Precipitation of the Group

$$Pb^{2+}(aq) + H_2S(aq) \xrightarrow{H^+(0.3\ M)} PbS(s,\ black) + 2\ H^+(aq) \qquad [35.18]$$

$$Cu^{2+}(aq) + H_2S(aq) \xrightarrow{H^+(0.3\ M)} CuS(s,\ black) + 2\ H^+(aq) \qquad [35.19]$$

$$Sn^{4+}(aq) + 2\ H_2S(aq) \xrightarrow{H^+(0.3\ M)} SnS_2(s,\ yellow) + 4\ H^+(aq) \qquad [35.20]$$

$$Hg^{2+}(aq) + H_2S(aq) \xrightarrow{H^+(0.3\ M)} HgS(s,\ black) + 2\ H^+(aq) \qquad [35.21]$$

$$2\ Bi^{3+}(aq) + 3\ H_2S(aq) \xrightarrow{H^+(0.3\ M)} Bi_2S_3(s,\ brown) + 6\ H^+(aq) \qquad [35.22]$$

Add about 10 to 12 drops of 1 M thioacetamide. Place the test tube into a boiling water bath (see Figure E, page 286) and heat for 10 min to saturate the solution thoroughly with $H_2S(aq)$ and precipitate all the metal sulfides that will separate out. Centrifuge the mixture and decant the clear, dark-colored solution carefully and completely into a clean test tube. Test the clear solution for completeness of precipitation by reheating it and adding 1 to 2 more drops of 1 M thioacetamide. Discard the centrifugate (S3A) unless the unknown is a general unknown containing ions from Groups IV and V. In the latter case label it as S3A and save for later analysis.

Wash the precipitate (R3) by adding about 4 mL H_2O to which 4 drops of 6 M HCl have been added. Stir this mixture well, centrifuge it, and discard the wash solution.

Separation of the Copper and Tin Subgroups

$$3\ SnS_2(s) + 6\ OH^- \longrightarrow [Sn(OH)_6]^{2-}(aq) + 2[SnS_3]^{2-}(aq) \qquad [35.23]$$

$$HgS(s) + HS^-(aq) + OH^-(aq) \longrightarrow [HgS_2]^{2-}(aq) + H_2O \qquad [35.24]$$

To R3, add 5 drops of 6 M NaOH and about 1 mL of water. Stir well and warm in the water bath for 2 to 3 min. Cool it, and add 5 drops of 3% hydrogen peroxide to oxidize any tin(II) ion to the tin(IV) state. Stir again and heat for about 5 min in the water bath until the effervescence caused by the decomposing H_2O_2 ceases (about 5 min). Any dark-colored precipitates present, such as CuS, PbS, Bi_2S_3, or HgS, would still remain. Now add 5 drops of 1 M thioacetamide and 5 drops of 6 M NaOH (this produces NaHS on heating). Stir, warm the solution (2 to 3 min in the water bath) and centrifuge it. Decant and save the solution (S4) for analysis of Hg^{2+} and Sn^{4+}. Wash any dark residue (R4) with 3 mL H_2O containing a drop each of 1 M thioacetamide and 6 M NaOH, stir well, centrifuge, and discard the washings.

The Copper Subgroup

$$3\ PbS + 8\ H^+(aq) + 2\ NO_3^-(aq) \longrightarrow 3\ Pb^{2+}(aq) + 3\ S(s) + 2\ NO(g) + 4\ H_2O \qquad [35.25]$$

$$3\ CuS + 8\ H^+(aq) + 2\ NO_3^-(aq) \longrightarrow 3\ Cu^{2+}(aq) + 3\ S(s) + 2\ NO(g) + 4\ H_2O \qquad [35.26]$$

$$Bi_2S_3 + 4\ H^+(aq) + 2\ NO_3^-(aq) \longrightarrow 2\ BiO^+(aq) + 3\ S(s) + 2\ NO(g) + 2\ H_2O \qquad [35.27]$$

$$Pb^{2+}(aq) + HSO_4^-(aq) \longrightarrow PbSO_4(s,\ white) + H^+(aq) \qquad [35.28]$$

To R4, add 1 mL of H_2O and 1 mL of 6 M HNO_3, warm, and stir the mixture well for several minutes to dissolve the sulfides. Centrifuge and decant the solution from the bit of dark residue (largely sulfur with a trace of undissolved sulfides) into a small beaker. Add 2 mL of 3 M H_2SO_4 to the solution, evaporate this until almost dry and very dense white choking fumes of SO_3 are evolved. (Use the hood!) The excessive escape of these fumes into the room can be prevented by covering the beaker with a watch glass after the evaporation. The purpose of this heating is to remove all HNO_3 in which lead sulfate is somewhat soluble. While still in the hood, cool the beaker, add 3 mL of H_2O, and stir well any residue remaining in the beaker. Remove from the hood, transfer to a small tube and centrifuge. Decant the solution (S5) and test it for copper and bismuth ions. Test any residue (R5) for lead.

Test for Lead Ion

$$PbSO_4(s) + 2\ C_2H_3O_2^-(aq) \longrightarrow Pb(C_2H_3O_2)_2(aq) + SO_4^{2-}(aq) \qquad [35.29]$$

$$Pb(C_2H_3O_2)_2(aq) + CrO_4^{2-}(aq) \longrightarrow PbCrO_4(s,\ yellow) + 2\ C_2H_3O_2^-(aq) \qquad [35.30]$$

To R5, add 1 mL of 3 M ammonium acetate, $NH_4C_2H_3O_2$, stir and warm if necessary, to dissolve any lead sulfate. Add 1 mL of 1 M K_2CrO_4 and centrifuge. A yellow precipitate proves that lead ion is present.

Test for Copper(II) Ion

$$Cu^{2+}(aq) + 4\,NH_3(aq) \longrightarrow [Cu(NH_3)_4]^{2+}(aq, \textit{deep blue}) \tag{35.31}$$

$$BiO^+(aq) + NH_3(aq) + H_2O \longrightarrow BiO(OH)(s, \textit{white}) + NH_4^+(aq) \tag{35.32}$$

To S5, add sufficient 6 M NH_3(aq) to make the solution distinctly basic. The appearance of a darker blue-violet color of $[Cu(NH_3)_4]^{2+}$ in the solution, S6, proves the presence of Cu^{2+}. Centrifuge if there is a precipitate, and save this (R6) for the bismuth test.

Test for Bismuth Ion

$$2\,BiO(OH)(s) + 3\,Sn(OH)_3^-(aq) + 3\,OH^-(aq) + 2\,H_2O \longrightarrow 2\,Bi(s, \textit{black}) + 3[Sn(OH)_6]^{2-}(aq) \tag{35.33}$$

To R6, add a drop each of 6 M NaOH and 0.1 M $SnCl_2$, stir well and centrifuge. The appearance of a gray-black color, Bi, in the residue proves that bismuth ion is present.

Precipitation of the Tin Subgroup

$$[SnS_3]^{2-}(aq) + 2\,H^+(aq) \longrightarrow SnS_2(s) + H_2S(aq) \tag{35.34}$$

$$[HgS_2]^{2-}(aq) + 2\,H^+(aq) \longrightarrow HgS(s) + H_2S(aq) \tag{35.35}$$

To the centrifugate (S4), which is strongly basic and contains S^{2-}, add enough 3 M H_2SO_4 drop by drop to neutralize the solution, stirring well after the addition of each drop. Test the solution by touching the stirring rod to litmus paper and note the total volume of acid used. Mix it well, but be cautious or the H_2S may be evolved too rapidly. (Work under the hood.) As long as H_2S is evolved rapidly on the addition of each drop of H_2SO_4, the solution is still basic. After the solution has been neutralized add an excess of 3 M H_2SO_4 equal to one-third the volume of H_2SO_4 needed for neutralization.* Centrifuge, decant, and discard the solution, S7, into the appropriate waste container. A dark or colored residue (R7) indicates the presence of the tin subgroup. If the residue is white or very pale yellow, it is sulfur, formed by the decomposition of some S^{2-} ion present.

---

*Since HSO_4^- is a weak acid ($K_a = 1.2 \times 10^{-2}$), a slight excess of H_2SO_4 reacts with some of the SO_4^{2-} formed

$$H_2SO_4 + SO_4^{2-} \longrightarrow 2\,HSO_4^-$$

so that the acidity is repressed by a buffering action. With one-third excess acid, the SO_4^{2-} and HSO_4^- concentrations will be about equal, so that

$$\frac{[H^+][SO_4^{2-}]}{[HSO_4^-]} = 1.2 \times 10^{-2}$$

$[H^+] = 1.2 \times 10^{-2}$ M.

This is not sufficient $[H^+]$ to dissolve the HgS, SnS_2 precipitate.

Separation of Mercury(II) Ion

$$3 SnS_2(s) + 6 OH^-(aq) \longrightarrow [Sn(OH)_6]^{2-}(aq) + 2 [SnS_3]^{2-}(aq) \qquad [35.36]$$

Wash R7 (possible HgS, SnS_2) with H_2O, stir well, centrifuge and discard the wash liquid. Then add about 1 mL of H_2O, 10 drops of 6 M NaOH, and 6 drops of 30% (not 3%) H_2O_2, and heat the mixture to decompose excess H_2O_2. (To avoid expulsion of material by too rapid steam formation, this heating may be carried out by immersing the test tube in boiling water in a small beaker, until evolution of $O_2(g)$ almost entirely ceases.) Centrifuge the mixture and decant and save the solution (S8) for the test for tin ion. Test any residue (R8) for mercury ion and the solution (S8), possibly containing $[Sn(OH)_6]^{2-}$, for tin ion. (*Note*: The S^{2-} formed by the solution of the sulfides, unless destroyed completely by oxidation with H_2O_2, might redissolve some HgS, which is otherwise insoluble in OH^- and thus render the separation ineffective.)

Test for Mercury(II) Ion

$$HgS(s) + 4 H^+(aq) + 2 NO_3^-(aq) + 4 Cl^-(aq) \longrightarrow [HgCl_4]^{2-}(aq) + S(s) + 2 NO_2(g) \qquad [35.37]$$

$$2 [HgCl_4]^{2-} + Sn^{2+} \longrightarrow Hg_2Cl_2(s) + [SnCl_6]^{2-}(aq) \qquad [35.38]$$

While stirring, wash R8 with H_2O, centrifuge, and decant the wash liquid. Then dissolve R8 by adding 5 drops each of 6 M HCl and 6 M HNO_3 and heat to boiling for 1 minute. After R8 has dissolved, add 1 mL of H_2O, then add 2 to 3 drops of 0.1 M $SnCl_2$ to the clear solution. A white silky precipitate of Hg_2Cl_2 turning dark with excess $SnCl_2$, indicates the presence of mercury(II) ion.

Test for Tin Ion

$$[Sn(OH)_6]^{2-}(aq) + 6 H^+(aq) + Fe(s) \longrightarrow Sn^{2+}(aq) + Fe^{2+}(aq) + 3 H_2O \qquad [35.39]$$

$$Sn^{2+}(aq) + 8 Cl^-(aq) + 2 Hg^{2+}(aq) \longrightarrow Hg_2Cl_2(s) + [SnCl_6]^{2-}(aq) \qquad [35.40]$$

Add 6 M HCl to S8 to neutralize the solution (litmus) and then 1 drop in excess. Evaporate to about 3 drops volume, then (but not before) add a one-half inch iron brad or 1 cm iron wire. Let this stand several minutes to reduce any tin ion to Sn^{2+}, then add 1 to 2 mL of H_2O and 2 drops of 0.1 M $HgCl_2$. A white silky precipitate of Hg_2Cl_2 (or possibly dark Hg if much tin ion is present) indicates the presence of tin ion.

DISPOSAL

Aqueous solutions: Dispose in a common waste bottle labeled for *qualitative analysis—solutions*.
Solids: Dispose in a waste bottle labeled *qualitative analysis—solids*.
Wash solutions/solids mixtures: Dispose in a waste bottle labeled *qualitative analysis—wash mixtures*.
All of these wastes will be collected, treated, and stored for volume reduction and proper disposal by qualified laboratory personnel.

Prelab   Name _____ Section _____

Group III: The Hydrogen Sulfide Group

1. If the initial metal sulfide precipitate is black with traces of yellow, what metal ion is likely to be present?

2. What metal sulfides are soluble in NaOH? How is this solubility used in this experiment?

3. In the section discussing the precipitation of the tin subgroup, why does the addition of sulfuric acid cause the precipitation of the mercury and tin metal sulfides?

4. Which metal ion in Group III was identified previously and why does it appear in two different groups?

5. What is the purpose of the $SnCl_2$ added to the solution formed by dissolving R8 in the test for mercury(II) ion?

Figure 35-0. Flowchart–The Hydrogen Sulfide Group

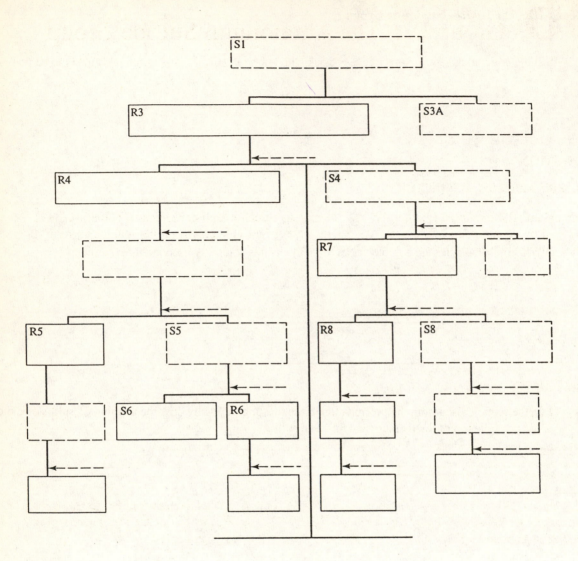

36 Group IV: The Ammonium Sulfide Group

INTRODUCTION

Precipitation with Ammonium Sulfide

An ammonium sulfide solution is rather difficult to keep, since such a high concentration of sulfide ion is readily oxidized by the air to free sulfur. Sulfur then dissolves in the ammonium sulfide solution to form a yellow ammonium polysulfide solution.

$$S^{2-}(aq) + S(s) \longrightarrow S_2^{2-}(aq) \qquad\qquad [36.1]$$

This latter ion gives a bothersome precipitate or colloid of free sulfur when the solution is acidified.

$$S_2^{2-}(aq) + 2\ H_3O^+(aq) \longrightarrow H_2S(g) + S(s) + 2\ H_2O(l) \qquad\qquad [36.2]$$

Therefore, we prepare the ammonium sulfide reagent as we use it, by making the test solution basic with $NH_3(aq)$ and then saturating it with $H_2S(aq)$.

$$2\ NH_3(aq) + H_2S(aq) \longrightarrow 2\ NH_4^+(aq) + S^{2-}(aq) \qquad\qquad [36.3]$$

Actually, the reaction does not proceed very much beyond the first stage of neutralization

$$NH_3(aq) + H_2S(aq) \longrightarrow NH_4^+(aq) + HS^-(aq) \qquad\qquad [36.4]$$

since both aqueous NH_3 and hydrogen sulfide are so poorly ionized that ammonium sulfide, $(NH_4)_2S$, is hydrolyzed largely to ammonium hydrogen sulfide, NH_4HS. The ionic equation for the precipitation of a metallic sulfide in ammonia solution is, therefore, principally (using Zn^{2+} as an example)

$$NH_3(aq) + Zn^{2+}(aq) + HS^-(aq) \longrightarrow ZnS(s) + NH_4^+(aq). \qquad\qquad [36.5]$$

However, for such reactions, we usually write the simpler equation

$$Zn^{2+}(aq) + S^{2-}(aq) \longrightarrow ZnS(s). \qquad\qquad [36.6]$$

The Hydrolysis of Sulfides of Weak Bases

The student will notice that two of the ions in the group are precipitated as hydroxides, not as sulfides, on the addition of ammonium sulfide solution. This is due to the large extent to which salts of weak bases and weak acids hydrolyze. Aluminum hydroxide is such a weak base that when ammonium sulfide is added to $Al^{3+}(aq)$ solution, we have the reactions

$$2\ Al^{3+}(aq) + 12\ H_2O \longrightarrow 2\ Al(OH)_3(s) + 6\ H_3O^+(aq)$$

$$3\ S^{2-}(aq) + 6\ H_2O \longrightarrow 3\ H_2S(g) + 6\ OH^-(aq)$$

which sum together to yield

$$2 \text{ Al}^{3+}(aq) + 3 \text{ S}^{2-}(aq) + 6 \text{ H}_2\text{O} \longrightarrow 2 \text{ Al(OH)}_3(s) + 3 \text{ H}_2\text{S}(g) \qquad [36.7]$$

The hydrolysis of both the aluminum ion and the sulfide ion is shifted completely to the right, since the products of the reaction, hydrogen ion and hydroxide ion, neutralize one another, and also because aluminum hydroxide separates as a precipitate and hydrogen sulfide is evolved as a gas. Chromium(III) ion, $Cr^{3+}(aq)$, behaves in a similar fashion.

Buffer Action

Frequently in chemical processes it is desirable to maintain the pH, or relative acidity and basicity of a solution, at an approximately constant value, even when substantial quantities of base or acid are added. Review the behavior of buffer solutions and the mechanisms by which this stability of pH control is attained, as described in your textbook.

In the precipitation of the ammonium sulfide group, aqueous ammonia is added to make the solution basic before saturating it with $H_2S(aq)$. It is important, however, that the solution not become too basic, or magnesium ion from the alkaline earth group may precipitate as magnesium hydroxide. Also, a strong aqueous ammonia solution has a hyroxide ion concentration almost high enough to re-dissolve aluminum hydroxide as the amphoteric hydroxo- complex ion, $[Al(OH)_4]^-$. These situations are avoided by having present a moderate concentration of ammonium ion (added as ammonium chloride) to maintain the hydroxide ion concentration at a value somewhat less than that of aqueous ammonia alone. The buffer action may be explained as follows.

In a mixture of aqueous ammonia and ammonium ion, with a strong acid added, the hydronium ions react with the aqueous ammonia present, or if a strong base is added, the hydroxide ions react with the ammonium ion present.

$$\text{NH}_3(aq) + \text{H}_3\text{O}^+(aq) \longrightarrow \text{NH}_4^+(aq) + \text{H}_2\text{O} \qquad [36.8]$$

$$\text{NH}_4^+(aq) + \text{OH}^-(aq) \longrightarrow \text{NH}_3(aq) + \text{H}_2\text{O} \qquad [36.9]$$

Thus the mixture does not change its pH greatly with the addition of either acid or base. This is further explained by the equilibrium expressions for the ionization of aqueous ammonia.

$$\text{H}_2\text{O} + \text{NH}_3(aq) \rightleftharpoons \text{NH}_4^+(aq) + \text{OH}^-(aq) \qquad [36.10]$$

$$\frac{[\text{NH}_4^+][\text{OH}^-]}{[\text{NH}_3(aq)]} = K_b = 1.74 \times 10^{-5} \qquad [36.11]$$

This can be written as

$$\frac{[\text{NH}_4^+]}{[\text{NH}_3(aq)]} \times [\text{OH}^-] = 1.74 \times 10^{-5} \qquad [36.12]$$

and thus we see that the hydroxide ion concentration depends on the ratio of the concentration of ammonium ion $[\text{NH}_4^+]$, to that of free aqueous ammonia $[\text{NH}_3(aq)]$. As long as there is a considerable amount of both salt and base present, the ratio $[\text{NH}_4^+]/[\text{NH}_3]$ will not change greatly on the addition of moderate amounts of acid or base. The hydroxide ion concentration will be correspondingly constant.

We shall also make use of another buffer, a mixture of sodium sulfate and sodium hydrogen sulfate. The hydrogen sulfate ion is a weak acid, as represented by its equilibrium expression.

$$\text{HSO}_4^-(aq) + \text{H}_2\text{O} \rightleftharpoons \text{H}_3\text{O}^+(aq) + \text{SO}_4^{2-}(aq) \qquad [36.13]$$

$$\frac{[H_3O^+][SO_4^{2-}]}{[HSO_4^-]} = K_a = 1.29 \times 10^{-2} \hspace{4cm} [36.14]$$

The addition of hydrogen ion shifts the above equilibrium to the left, and the addition of hydroxide ion shifts it to the right. However, as long as there is a considerable amount of both sulfate ion and hydrogen sulfate ion present, the ratio $[SO_4^{2-}]/[HSO_4^-]$ does not change much on addition of acid or base, and the hydrogen ion concentration, as shown by equation [36.14], should remain about 10^y M.

We have used this buffer previously to prevent the solution of the tin(IV) sulfide formed when $[SnS_3]^{2-}$ is acidified. We shall use it here to prevent the solution of zinc sulfide while dissolving all other sulfides and hydroxides of the group.

Oxidation States of Chromium and Manganese

While chromium and manganese are metals, they readily coordinate with oxygen to form oxides or negative complex ions when they have high oxidation states. Review the principal oxidation states of chromium and manganese, and the amphoteric character of chromium(III) hydroxide. Also, review the chart below on the oxidation states of oxygen so that you understand the behavior of hydrogen peroxide, H_2O_2.

		Acidic	**Basic**	
↑	0	O_2		│
Oxidized	−1	H_2O_2	HO_2^-	Reduced
│	−2	H_2O	OH^-	↓

We shall use H_2O_2 both as a good oxidizing agent, and, under different conditions, as a reducing agent.

We make use of the change in properties of chromium and manganese ions with change in oxidation state in order to separate and test for these elements. The properties of chromium(III) ion (Cr^{3+}) are very similar to those of the trivalent ions of aluminum (Al^{3+}) and iron (Fe^{3+}). However, when the chromium(III) ion is oxidized to chromate ion (CrO_4^{2-}), it behaves very differently and is readily separated and tested. The oxidation is accomplished by hydrogen peroxide in a basic solution. The peroxide ion in basic solution becomes HO_2^-. The excess hydrogen peroxide is readily decomposed into water and oxygen gas by boiling the solution. (Note that no oxygen gas is formed, however, by the hydrogen peroxide that oxidizes the chromium(III) ion, as will be shown in equation [36.34]).

Any manganese(II) ion present in the above treatment with hydrogen peroxide in basic solution is oxidized and precipitated as manganese dioxide [MnO_2, or more probably a hydrated form $MnO(OH)_2$]. After the manganese dioxide precipitate is separated, it is redissolved by hydrogen peroxide in a nitric acid solution. In this case, the manganese dioxide is reduced to manganese(II) ion by the hydrogen peroxide, which in turn is oxidized to oxygen. Finally, the very characteristic purple permanganate ion (MnO_4^-) is produced by adding the powerful oxidizing agent, sodium bismuthate ($NaBiO_3$), which is reduced to the trivalent bismuth ion (Bi^{3+}). These changes in the oxidation state of manganese are indicated in Figure 36-1.

SAFETY PRECAUTIONS

Review the safety rules on pages 1 and 2.

Wear safety goggles at all times.

Be particularly careful when heating and mixing chemicals.

Use the hood to exhaust any fumes that might be released when carrying out the following reactions.

Use particular care when using 30% H_2O_2. If any gets on the skin, wash immediately with water for at least 5 min.

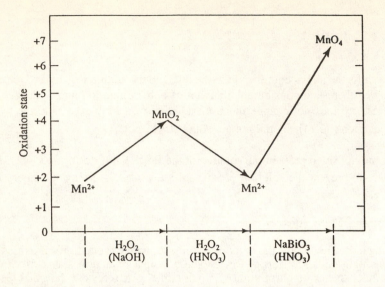

Figure 36-1. Changes in Oxidation State in the Separation of Manganese

PROCEDURE FOR THE ANALYSIS OF AN UNKNOWN SOLUTION FOR Fe^{3+}, Al^{3+}, Cr^{3+}, Mn^{2+}, Zn^{2+}, Ni

Before starting analysis, complete the ammonium sulfide group flowchart (Figure 36-2).

$$Fe^{3+}(aq) + 3\ NH_3(aq) + 3\ H_2O \longrightarrow Fe(OH)_3\ (s, \textit{red-brown})r + 3\ NH_4^+(aq) \qquad [36.15]$$

$$Al^{3+}(aq) + 3\ NH_3(aq) + 3\ H_2O \longrightarrow Al(OH)_3\ (s, \textit{white}) + 3\ NH_4^+(aq) \qquad [36.16]$$

$$Cr^{3+}(aq) + 3\ NH_3(aq) + 3\ H_2O \longrightarrow Cr(OH)_3\ (s, \textit{green}) + 3\ NH_4^+(aq) \qquad [36.17]$$

$$Mn^{2+}(aq) + NH_3(aq) + HS^-(aq) \longrightarrow MnS\ (s, \textit{pale pink}) + NH_4^+(aq) \qquad [36.18]$$

$$Zn^{2+}(aq) + NH_3(aq) + HS^-(aq) \longrightarrow ZnS\ (s, \textit{white}) + NH_4^+(aq) \qquad [36.19]$$

$$Ni^{2+}(aq) + NH_3(aq) + HS^-(aq) \longrightarrow NiS\ (s, \textit{black}) + NH_4^+(aq) \qquad [36.20]$$

$$Fe^{2+}(aq) + NH_3(aq) + HS^-(aq) \longrightarrow FeS\ (s, \textit{black}) + NH_4^+(aq) \qquad [36.21]$$

Add 1 mL of 3 M NH_4Cl to 3 mL of the test solution*, and then add 6 M $NH_3(aq)$ to neutralize the mixture, and about 0.5 mL in excess. The absence of a precipitate here (it will flocculate more on standing for several minutes) indicates the absence of Fe^{3+}, Al^{3+}, and Cr^{3+}. Without centrifuging any precipitate, saturate the solution thoroughly with $H_2S(aq)$, by adding 10 to 12 drops of 1 M thioacetamide, stirring well, and heating for 5 min in the water bath. Again, add 0.5 mL of 6 M $NH_3(aq)$, mix well, and warm slightly. Centrifuge the mixture. (Divide into two tubes to centrifuge if there is too large a volume for one tube.) Save the solution (S9), acidifying it with HCl and heating for 5 min in the boiling water bath, if analyses for Group V are to be made. Otherwise, dispose of it in the labeled waste container. Wash the residue (R9) with a wash solution made by adding 2 drops each of   6 M $NH_3(aq)$ and 3 M NH_4Cl to 1 mL of water. Centrifuge again and discard the wash liquid. Do not let R9 dry; proceed at once to the separation of ZnS and NiS.

---

*The test solution may be either S3A from the separation of Group III or a solution limited to Group IV ions. In either case it is likely to be acidic and will need to be neutralized prior to analysis of Group IV ions.

Separation of ZnS and NiS

$$Fe(OH)_3(s) + 3\ H_3O^+(aq) \longrightarrow Fe^{3+}(aq) + 6\ H_2O \qquad\qquad [36.22]$$

$$Al(OH)_3(s) + 3\ H_3O^+(aq) \longrightarrow Al^{3+}(aq) + 6\ H_2O \qquad\qquad [36.23]$$

$$Cr(OH)_3(s) + 3\ H_3O^+(aq) \longrightarrow Cr^{3+}(aq) + 6\ H_2O \qquad\qquad [36.24]$$

$$FeS(s) + 2\ H_3O^+(aq) \longrightarrow Fe^{2+}(aq) + H_2S(aq) + 2\ H_2O \qquad\qquad [36.25]$$

$$MnS(s) + 2\ H_3O^+(aq) \longrightarrow Mn^{2+}(aq) + H_2S(aq) + 2\ H_2O \qquad\qquad [36.26]$$

To R9, first add 2 mL of 1 M Na_2SO_4 and then 2 mL of 1 M $NaHSO_4$. Stir this mixture for 1 min. (Do not heat it.) Centrifuge, decant S10, and test the residue R10 for the analysis of zinc and nickel ions. Heat S10 in the boiling water bath for at least 10 min to remove all H_2S gas, cool and save for the analysis of iron, manganese, aluminum and chromium ions.

Test for Zinc Ion

$$ZnS(s) + 2\ H_3O^+(aq) \longrightarrow Zn^{2+}(aq) + H_2S(aq) + 2\ H_2O \qquad\qquad [36.27]$$

$$Zn^{2+}(aq) + HS^-(aq) + H_2O \longrightarrow ZnS(s) + H_3O^+(aq) \qquad\qquad [36.28]$$

$$3\ Zn^{2+}(aq) + 2\ K^+(aq) + 2[Fe(CN)_6]^{4-}(aq) \longrightarrow K_2Zn_3[Fe(CN)_6]_2\ (s,\ gray) \qquad\qquad [36.29]$$

$$K_2Zn_3[Fe(CN)_6]_2(s) + 12\ OH^-(aq) \longrightarrow 2\ K^+(aq) + 3[Zn(OH)_4]^{2-}(aq) + 2[Fe(CN)_6]^{4-}(aq) \qquad [36.30]$$

Wash the above residue (R10)* with 5 mL of H_2O to remove the buffer mixture. (If the precipitate tends to be colloidal, add 2 drops of 3 M NH_4Cl.) Centrifuge this and discard the wash solution. Add 1 mL of H_2O to the residue, and then add 2 drops of 6 M HCl. Stir the mixture for 1 min, then centrifuge it. [Save the residue (R11) for the analysis of Ni^{2+}.] Add 1 drop of 1M thioacetamide and 1 drop of 6 M NaOH to solution (S11), and heat for 5 min. This produces NaHS, which insures sufficient sulfide ion. Add 3 M $NH_4C_2H_3O_2$ by drops, up to 10 drops, while stirring. Centrifuge and discard the wash liquid in the labeled waste container. A white or light gray precipitate of ZnS indicates zinc ion. Add 10 drops of water to the precipitate and then add 6 M HCl dropwise with stirring until the mixture is just acidic. (Test with blue litmus after the addition of each drop.) If the precipitate has not completely dissolved, centrifuge the mixture and separate the clear solution. To this solution, add 10 to 20 drops of 0.1 M $K_4[Fe(CN)_6]$ and stir well. If there is no precipitate, zinc ion is absent. If a precipitate forms, separate it from the solution and attempt to dissolve it in 12 drops of 6 M NaOH. Stir the mixture thoroughly, while heating it in a water bath. If the precipitate dissolves completely (reaction [36.30]), the presence of zinc ion is confirmed.

---

*A light gray color may indicate ZnS. If it is quite dark, it may be due to nickel sulfide.

Test for Ni²⁺ Ion

$$NiS(s) + 4 H_3O^+(aq) + 2 NO_3^-(aq) \longrightarrow Ni^{2+}(aq) + 2 NO_2(g) + 6 H_2O(l) + S(s) \qquad [36.31]$$

$$Ni^{2+}(aq) + 2 HDMG + 2 NH_3(aq) \longrightarrow Ni(DMG)_2 (s,\ red) + 2 NH_4^+(aq) \qquad [36.32]$$

Add 2 drops of 6 M HNO_3 to R11. Warm this to dissolve the sulfides, and evaporate the solution to 1 drop. Add 1 mL of H_2O. (The dark-colored residual sulfur may be removed with a stirring rod.)

Add 6 M NH_3(aq) to make the solution basic, and then add 2 to 3 drops of 1% dimethylglyoxime. A red precipitate, $NiC_8H_{14}N_4O_4$ or $Ni(DMG)_2$ (rxn.36.32), confirms the presence of nickel(II) ion.

Separation of the Iron and Aluminum Subgroups

$$Al^{3+}(aq) + 3 OH^-(aq) \longrightarrow [Al(OH)_4]^-(aq) \qquad [36.33]$$

$$2 Cr^{3+}(aq) + 3 H_2O_2(aq) + 10 OH^-(aq) \longrightarrow 2 CrO_4^{2-}(aq) + 8 H_2O \qquad [36.34]$$

$$Mn^{2+}(aq) + H_2O_2(aq) + 2 OH^-(aq) \longrightarrow MnO_2(s) + 2 H_2O \qquad [36.35]$$

$$2 Fe^{2+}(aq) + H_2O_2(aq) + 4 OH^-(aq) \longrightarrow 2 Fe(OH)_3(s) \qquad [36.36]$$

$$Fe^{3+}(aq) + 3 OH^-(aq) \longrightarrow Fe(OH)_3(s) \qquad [36.37]$$

Add 5 drops of 30% H_2O_2 to S10, stir and add this mixture to a 15 cm test tube containing 1 mL of H_2O and 1 mL of 6 M NaOH. The order of addition of reagents is important as it ensures an adequate mixing of reagents and opportunity for complete oxidation before the H_2O_2 is decomposed by the basic solution. Failure to obtain a good test for chromium ion is often due to incomplete oxidation of Cr^{3+} to CrO_4^{2-}. Mix the solution and let it stand for 1 min, then heat it cautiously to decompose the H_2O_2 completely (as evidenced by no further effervescence of O_2 when the heat is withdrawn). Finally, centrifuge the mixture and save the solution (S12) for the tests for aluminum and chromium ions. Test the residue (R12) for iron ion.

Test for Iron (Fe³⁺)

$$Fe(OH)_3(s) + 3 H_3O^+(aq) \longrightarrow Fe^{3+}(aq) + 6 H_2O \qquad [36.38]$$

$$MnO_2(s) + H_2O_2(aq) \longrightarrow Mn^{2+}(aq) + O_2(g) + 2 OH^-(aq) \qquad [36.39]$$

$$Fe^{3+}(aq) + SCN^-(aq) \longrightarrow Fe(SCN)^{2+} (aq,\ red) \qquad [36.40]$$

Treat R12 with 1 mL of 6 M HCl. If the precipitate does not dissolve completely,* add 2 drops of 3% H_2O_2 to reduce any MnO_2 to Mn^{2+}, and boil the solution to decompose excess H_2O_2.

---

*A separation of the Fe^{3+} solution and the MnO_2 precipitate can be made here, but it is unnecessary, as the tests for Fe^{3+} and for Mn^{2+} do not interfere with each other.

Cool the solution, dilute it to 2 mL with water, mixing well, and divide it into two portions. To one, add 6 M NH_3(aq) to neutralize excess acid, and then 6 M HCl dropwise to make it slightly acid again. Add up to 1 mL of 0.1 M KSCN (potassium thiocyanate) dropwise. A deep red color of $FeSCN^{2+}$ in solution (rxn. 36.40) confirms the presence of iron(III) ion.

Do not report iron ion if only a slight coloration results. This coloration may be due to impurities of iron ion in the reagents used, or possibly to rust particles carelessly introduced into the sample from test tube clamps, wire gauze, etc. If in doubt, test a sample of the original unknown with 0.1 M KSCN.

Test for Manganese(II) Ion

$$2\ Mn^{2+}(aq) + 5\ NaBiO_3(s) + 14\ H_3O^+(aq) \longrightarrow$$
$$2\ MnO_4^-(aq,\ purple) + 5\ Bi^{3+}(aq) + 21\ H_2O(l) + 5\ Na^+(aq) \qquad [36.41]$$

Place the other portion of the above solution in a hot water bath. After one minute, add a small amount (about a fourth of the volume of a pea) of solid sodium bismuthate ($NaBiO_3$). Stir constantly for 1 min and centrifuge. A pink or purple color of MnO_4^- in the solution (reaction [36.41]) confirms the presence of manganese(II) ion.

Test for Aluminum Ion

$$[Al(OH)_4]^-(aq) + 4\ H_3O^+(aq) \longrightarrow Al^{3+}(aq) + 8\ H_2O \qquad\qquad [36.42]$$

$$2\ CrO_4^{2-}\ (aq,\ yellow) + 2\ H_3O^+(aq) \longrightarrow Cr_2O_7^{2-}\ (aq,\ orange) + 3\ H_2O \qquad [36.43]$$

$$Al^{3+}(aq) + 3\ NH_3(aq) + 3\ H_2O \longrightarrow Al(OH)_3\ (s,\ white) + 3\ NH_4^+(aq) \qquad [36.44]$$

$$Al^{3+}(aq) + 3\ NH_3(aq) + 3\ H_2O \xrightarrow{\text{aluminon dye}} Al(OH)_3 \cdot aluminon\ (s,\ red) + 3\ NH_4^+(aq) \qquad [36.45]$$

While stirring acidify S12 with 6 M HCl (test by touching the stirring rod to litmus paper), and add about 5 drops in excess. Then make the solution just basic with 6 M NH_3(aq) (test with litmus paper). A flocculent precipitate indicates aluminum ion. Centrifuge and save the solution (S13) for the test for chromium ion. Wash the residue (R13) with 10 drops of distilled water and discard the wash liquid in the labeled waste container.

To confirm, dissolve R13 in 1 mL of 6 M HCl, add 2 drops of aluminon reagent, and then make the solution faintly basic by addition of 6 M NH_3(aq), drop by drop. A red precipitate, which flocculates as the mixture stands and is suspended in an otherwise colorless solution, indicates the presence of aluminum ion (reaction [36.45]). If in doubt, observe the test tube against a white background. If a good red color, which is completely absorbed in the precipitate, is not obtained, the solution possibly is too basic. In this case, reacidify with HCl, and again make just basic with NH_3(aq).

This precipitate is quite characteristic. Silica, often present as an impurity from the NaOH used, gives a white precipitate. $Cr(OH)_3$, which might be present from unoxidized Cr^{3+}, gives a precipitate similar to that of $Al(OH)_3$, but it redissolves on the addition of 0.5 mL of 3 M $(NH_4)_2CO_3$. Any Fe^{2+} that failed to be completely separated from S10 would give a reddish color to the original $Al(OH)_3$ precipitate before the aluminon reagent is added. In this case, treat this precipitate with a little 6 M NaOH, centrifuge the undissolved $Fe(OH)_3$, and acidify the solution with 6 M HCl. Then follow the aluminon reagent treatment as described above.

Test for Chromate Ion

$$CrO_4^{2-}(aq) + Pb(C_2H_3O_2)_2(aq) \longrightarrow PbCrO_4 (s, \textit{yellow}) + 2\ C_2H_3O_2^-(aq) \qquad [36.46]$$

Add 1 M $HC_2H_3O_2$ to S13 to make it slightly acidic, and then add 5 to 10 drops of 0.1 M $Pb(C_2H_3O_2)_2$. A yellow precipitate of $PbCrO_4$ confirms chromium ion (reaction [36.46]). If necessary, the mixture may be centrifuged to obtain a better observation of the color of the precipitate. Avoid a large excess of $Pb(C_2H_3O_2)_2$. A white precipitate of $PbSO_4$ will form in the absence of $PbCrO_4$, but the latter is much less soluble and will precipitate first.

DISPOSAL

Aqueous solutions: Dispose in a common waste bottle labeled for *qualitative analysis—solutions*.

Solids: Dispose in a waste bottle labeled *qualitative analysis—solids*.

Wash solutions/solids mixtures: Dispose in a waste bottle labeled *qualitative analysis—wash mixtures*.

All of these wastes will be collected, treated, and stored for volume reduction and proper disposal by qualified laboratory personnel.

Prelab   Name _____ Section _____

Group IV: The Ammonium Sulfide Group

1. If a green precipitate is formed by the addition of NH_4Cl, NH_3, and thioacetamide at the beginning of the experiment, what ion is probably present? How is the presence of this ion confirmed?

2. Which three metal cations are precipitated initially as hydroxides rather than sulfides?

3. Write a chemical equation that describes the use of H_2O_2 as an oxidizing agent in this experiment. Write a second equation showing H_2O_2 as a reducing agent.

4. Under the procedure heading, Test for Zinc Ion, residue R10 is washed and then acidified with HCl. If all of R10 dissolves on addition of the acid, what ion is probably present and what other ion is probably absent?

5. If you have a solid precipitate, R12, what two ions are possibly present?

Figure 36-2. Flowchart: The Ammonium Sulfide Group

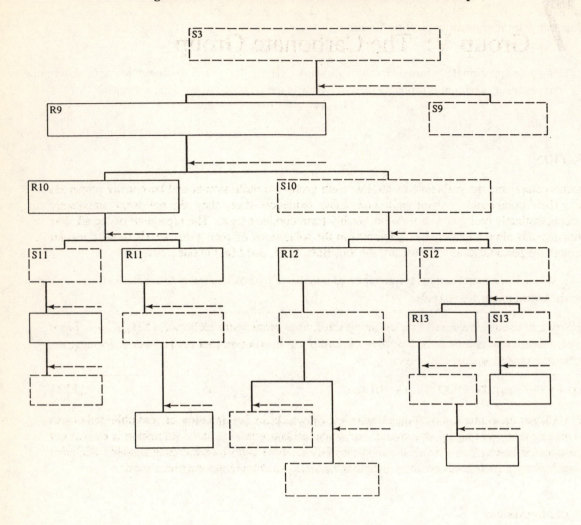

37 Group V: The Carbonate Group

INTRODUCTION

The elements comprising the carbonate or alkaline earth group are quite similar and have many properties in common. Their compounds exhibit only one stable oxidation state; they do not form amphoteric hydroxides, being distinctly basic; and they do not readily form complex ions. The separation of the alkaline earth elements depends almost entirely on differences in the solubilities of their salts, which show a regular gradation through the periodic table. We include the ions Ba^{2+}, Ca^{2+}, and Mg^{2+} in our study.

Precipitation with Ammonium Carbonate

The group reagent used to precipitate the alkaline earth ions is ammonium carbonate, $(NH_4)_2CO_3$. Since this is a salt of a weak base and of a weak acid, it will hydrolyze in solution to a marked extent, forming the hydrogen carbonate ion and aqueous ammonia.

$$NH_4^+(aq) + CO_3^{2-}(aq) \leftrightarrows HCO_3^-(aq) + NH_3(aq) \qquad [37.1]$$

This results in a lower concentration of the carbonate ion, on which the precipitation of insoluble carbonates depends. To prevent this, and thereby increase the carbonate ion concentration, the precipitation is carried out in aqueous ammonia solution. The hydrogen carbonates of the alkaline earth ions are quite soluble. (Calcium and magnesium hydrogen carbonates are the principal substances present in temporary hard water.)

The Separation of Magnesium

Magnesium carbonate, being fairly soluble, is not precipitated readily unless the solution is concentrated and the carbonate ion concentration is very high. Furthermore, it tends to form supersaturated solutions and does not precipitate promptly. Since magnesium hydroxide is even less soluble than the carbonate, the magnesium ion might precipitate as the hydroxide when the ammonium carbonate and aqueous ammonia reagent is added. This can be prevented by buffering the solution with a high concentration of ammonium ion to reduce the hydroxide ion concentration. In the analysis of a general unknown, sufficient ammonium chloride ion for this purpose will be present from the previous treatment of the solution in the separation of the Group IV ions.

Although in some schemes of analysis all excess ammonium ion is removed and the carbonate ion concentration is made as high as possible to precipitate magnesium ion with the alkaline earth group, we shall leave it in solution.

The Separation of Barium, Calcium, and Magnesium

Compare the solubilities of these ions together with the following procedure for the analysis, and your flowchart. Note that, after the solution of the carbonate precipitate (R15) with acetic acid, barium can be separated effectively as the chromate, $BaCrO_4$, and then further identified by conversion to the white sulfate, $BaSO_4$. With other ions of the group removed, calcium is easily identified by precipitation with oxalate ion, as the quite insoluble CaC_2O_4.

PROCEDURE FOR THE ANALYSIS OF AN UNKNOWN SOLUTION FOR Ca^{2+}, Ba^{2+}, and Mg^{2+}

Before starting analysis, complete the carbonate group flowchart (Figure 37-0).

Precipitation of the Carbonate Group

$$Ba^{2+}(aq) + CO_3{}^{2-}(aq) \xrightarrow{\ NH_4{}^+(aq)\ } BaCO_3 \text{ (s, }white\text{)} \qquad\qquad [37.2]$$

$$Ca^{2+}(aq) + CO_3{}^{2-}(aq) \xrightarrow{\ NH_4{}^+(aq)\ } CaCO_3 \text{ (s, }white\text{)} \qquad\qquad [37.3]$$

If the unknown is being analyzed for this group only, to a 3 mL sample of it add 1 mL of 3 M NH_4Cl, and warm it almost to boiling. If the sample is S9 from a general unknown analysis, evaporate it to 3 to 4 mL but omit the addition of ammonium chloride as this is already present. To either of these hot samples, add 3 M $(NH_4)_2CO_3$ reagent by drops until the mixture is just basic, then 1 mL in excess. Mix this well, and let it stand for 5 min to complete any precipitation. Centrifuge. Save the solution (S14) for the analysis of magnesium ion and the residue, R14, for the analysis of Ba^{2+} and Ca^{2+}.

Test for Barium Ion

$$BaCO_3(s) + HC_2H_3O_2(aq) + NH_4{}^+(aq) \longrightarrow Ba^{2+}(aq) + H_2O(l) + CO_2(g) + NH_3(aq) + C_2H_3O_2{}^-(aq) \quad [37.4]$$

$$CaCO_3(s) + HC_2H_3O_2(aq) + NH_4{}^+(aq) \longrightarrow Ca^{2+}(aq) + H_2O(l) + CO_2(g) + NH_3(aq) + C_2H_3O_2{}^-(aq) \quad [37.5]$$

$$Ba^{2+}(aq) + CrO_4{}^{2-}(aq) \longrightarrow BaCrO_4 \text{ (s, }yellow\text{)} \qquad\qquad [37.6]$$

$$BaCrO_4(s) + H_3O^+(aq) \longrightarrow Ba^{2+}(aq) + HCrO_4{}^-(aq) + H_2O \qquad\qquad [37.7]$$

$$Ba^{2+}(aq) + HSO_4{}^-(aq) \longrightarrow BaSO_4 \text{ (s, }white\text{)} \qquad\qquad [37.8]$$

Wash R14 with 1 mL H_2O and discard the washings. Add 5 drops of 6 M $HC_2H_3O_2$ to the residue, warm the mixture gently to dissolve the carbonates, add 5 drops of 3 M $NH_4C_2H_3O_2$, and dilute the solution with water to 1 mL. Add 3 drops of 1 M K_2CrO_4, warm and agitate the mixture. Centrifuge. If the solution is not yellow, add more 1 M K_2CrO_4 by drops to complete the precipitation of $BaCrO_4$, but avoid over 3 drops in excess. Again centrifuge and save the solution (S15) for the Ca^{2+} test. To confirm the presence of barium ion, add 2 or 3 drops of 6 M HCl to dissolve the yellow precipitate (R15), dilute to 2 mL with water, and add 1 or 2 drops of 3 M H_2SO_4. A white precipitate of $BaSO_4$ (reaction [37.8]) confirms the presence of barium ion. (It may be necessary to centrifuge in order to see whether the precipitate is white in the yellow solution.)

Test for Calcium Ion

$$Ca^{2+}(aq) + C_2O_4{}^{2-}(aq) \xrightarrow{\ NH_4{}^+(aq)\ } CaC_2O_4 \text{ (s, }white\text{)} \qquad\qquad [37.9]$$

To S15 add 10 drops of 1 M $K_2C_2O_4$, and add 6 M $NH_3(aq)$ by drops until the solution is basic. (Orange $Cr_2O_7{}^{2-}$ changes to yellow $CrO_4{}^{2-}$.) (*Note*: $K_2C_2O_4$ is potassium oxalate, and its solution is colorless, whereas K_2CrO_4 is potassium chromate and its solution is yellow.) Let stand 10 min if a precipitate does not appear before that time. The formation of white insoluble CaC_2O_4 (reaction [37.9]) indicates calcium ion. It may be necessary to centrifuge to show that the precipitate is white in the yellow solution. Further confirmation may be made, if desired, by decanting the yellow solution and adding to the residue a few drops of 6 M HCl and 1 mL of H_2O to dissolve it. Then add a drop of 1 M $K_2C_2O_4$ and again make basic with $NH_3(aq)$. The white precipitate of CaC_2O_4 will reappear.

Test for Magnesium Ion

$$Mg^{2+}(aq) + 2\ NH_3(aq) + H_2PO_4^- + 6\ H_2O \longrightarrow MgNH_4PO_4{\cdot}6\ H_2O(s) + NH_4^+(aq) \qquad [37.10]$$

$$MgNH_4PO_4(s) + 2\ H_3O^+(aq) \longrightarrow Mg^{2+}(aq) + NH_4^+(aq) + H_2PO_4^-(aq) + 2\ H_2O \qquad [37.11]$$

$$Mg^{2+} + 2\ OH^-(aq) \xrightarrow[\text{reagent}]{\text{Mg}} Mg(OH)_2{-}(\text{paranitrobenzeneazoresorcinol})\ (s,\ blue) \qquad [37.12]$$

To S14, add 1 mL of 6 M NH_3(aq) and then 0.5 mL of 1 M NaH_2PO_4. Stir this at intervals for 5-10 min. A white precipitate of magnesium ammonium phosphate hexahydrate ($MgNH_4PO_4{\cdot}6\ H_2O$) is a preliminary indicator of magnesium. (If Ba^{2+} and Ca^{2+} were not removed completely before, they would give insoluble phosphate precipitates here.) Let the solution stand at least one-half hour before deciding that Mg^{2+} is absent, as this precipitate is often slow in forming. To confirm magnesium, centrifuge the mixture, wash the precipitate with water (discarding both solution and wash water), and dissolve it by adding several drops of 6 M HCl and 1 mL of H_2O. Add 1 drop of "magnesium reagent" (*p*-nitrobenzeneazoresorcinol) and make the solution alkaline with 6 M NaOH. Mix, and let stand for 1 to 5 min. Observe against good light for the characteristic blue lake (reaction [37.12]). $Mg(OH)_2$ is a flocculent adsorption compound with the dye adsorbed to its surface.

DISPOSAL

Aqueous solutions: Dispose in a common waste bottle labeled for *qualitative analysis—solutions*.
Solids: Dispose in a waste bottle labeled *qualitative analysis—solids*.
Wash solutions/solids mixtures: Dispose in a waste bottle labeled *qualitative analysis—wash mixtures*.
All of these wastes will be collected, treated, and stored for volume reduction and proper disposal by qualified laboratory personnel.

Prelab   Name _____ Section _____

Group V: The Carbonate Group

1. How is the Ba^{2+} separated from the Ca^{2+}?

2. What test is used to confirm the presence of calcium?

3. If small amounts of Ba^{2+} and Ca^{2+} remain in solution S14, how will the test for Mg^{2+} be affected?

4. How is the pH of the precipitating solution controlled?

5. Which of the following is most soluble: $MgCO_3$, $CaCO_3$, or $BaCO_3$?

Figure 37-0. Flowchart: The Carbonate Group

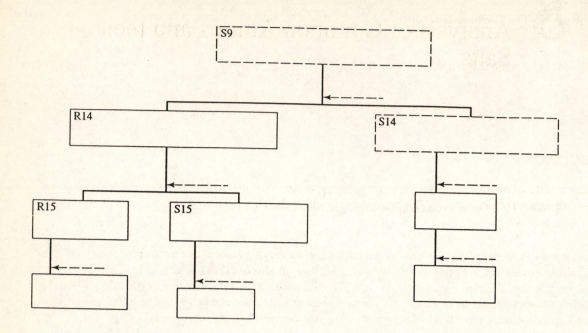

38 Analysis of Common Anions and their Salts

INTRODUCTION

To analyze salts or mixtures of salts we must apply procedures for the cation analysis, and in addition we must detect all anions (negative ions). This experiment involves the anions NO_3^-, NO_2^-, Cl^-, Br^-, I^-, SO_3^{2-}, SO_4^{2-}, CO_3^{2-}, and PO_4^{3-}.

Although there is no orderly procedure for separating and identifying anions, as there is for cations, definite steps can be taken to test for or prove the absence of each ion. In almost all cases several ions can be proved absent by a single test, thereby eliminating the need to make individual tests for those ions already proved absent. Furthermore, there are not many common anions and most of them can be identified in the presence of other anions; thus elaborate procedures for separating the anions from each other are unnecessary.

The common anions can be divided into three groups.

1. The silver group, which forms salts with silver ions that are insoluble in dilute nitric acid. The ions of this group selected for analysis are chloride (Cl^-), bromide (Br^-), and iodide (I^-).

2. The barium group, which forms salts with barium ions that, except for $BaSO_4$, are soluble in dilute nitric acid but insoluble in water. The ions selected from this group for analysis are sulfate (SO_4^{2-}), sulfite (SO_3^{2-}), phosphate (PO_4^{3-}), and carbonate (CO_3^{2-}).

3. The nitrate group, which does not form insoluble salts with either silver or barium ions. These anions are nitrate (NO_3^-) and nitrite (NO_2^-). It is called the nitrate group because the nitrate ion is the more common ion of this pair.

Analysis of a Sample

Before beginning the analysis of a salt or mixture of salts it is necessary to have a plan or outline of steps to take and the proper order in which to take them. The following plan is recommended.

1. Analyze half of the sample for the anions according to the following procedures.

2. Analyze the other half of the sample for the cations in the following order.

    a. Use a small portion to test for the presence of ammonium ion (NH_4^+), according to the procedure of Experiment 33.

    b. Test for Na^+ with a flame test (Experiment 33).

    c. Analyze the remaining solution, starting with Group IV. No cations are included from Groups II and III.

Inspection of a Sample

If the sample is colored, the color reveals something of the nature of the ions present. If the sample is colorless, all colored ions are absent and need not be tested for. Certain ions are colorless when dry, like the copper ion in anhydrous copper sulfate, but acquire their characteristic color when water is added. Inspect the sample to see if it is homogeneous and identify any and all colors. Do not rely entirely on colors seen for identification of ions and compounds. Always confirm your preliminary identification with proper chemical tests.

PROCEDURE

The following preliminary tests (the silver and barium group tests) are used to determine which anions are absent from your sample. It is not necessary to perform the test for individual anions that have been determined not to be present. Since neither silver nor barium precipitate nitrate or nitrite, the nitrate and nitrite test should be performed on all samples.

Solution of the Sample

Place an amount of sample the size of 2 or 3 grains of rice in a small test tube. Add about 2 mL of distilled water and stir to dissolve. The sample should dissolve readily; if it does not, heat in a water bath for a few minutes. The sample test solution is now ready for analysis. A freshly prepared test solution is needed for each of the following preliminary and most individual tests.

Preliminary Classification

The Silver Group Test

$$2\ Ag^+(aq) + CO_3^{2-}(aq) \longrightarrow Ag_2CO_3\ (s,\ white) \tag{38.1}$$

$$2\ Ag^+(aq) + SO_3^{2-}(aq) \longrightarrow Ag_2SO_3\ (s,\ cream\text{-}colored) \tag{38.2}$$

$$3\ Ag^+(aq) + PO_4^{2-}(aq) \longrightarrow Ag_3PO_4\ (s,\ yellow) \tag{38.3}$$

$$Ag^+(aq) + Cl^-(aq) \longrightarrow AgCl\ (s,\ white) \tag{38.4}$$

$$Ag^+(aq) + Br^-(aq) \longrightarrow AgBr\ (s,\ cream\text{-}colored) \tag{38.5}$$

$$Ag^+(aq) + I^-(aq) \longrightarrow AgI\ (s,\ light\text{-}yellow) \tag{38.6}$$

Add 2 drops of 0.1 M $AgNO_3$ solution to 5 drops of a test solution of the sample dissolved in water. The water-insoluble silver salts are silver carbonate, Ag_2CO_3; silver sulfite, Ag_2SO_3; silver chloride, AgCl; silver bromide, AgBr; silver iodide, AgI (light-yellow); and silver phosphate, Ag_3PO_4 (yellow). Centrifuge the precipitate and discard the solution in the labeled waste container. Add 5 drops of distilled water to the solid. Stir, centrifuge, and discard the water. Now add 5 drops of 3 M HNO_3 solution to the solid and observe it closely. If it all dissolves, neither chloride, bromide, nor iodide was present and individual tests for carbonate, sulfite, and phosphate are indicated. If not all the precipitate dissolves, one or more of these three ions is present in the sample. The color of the solid will give an indication of which ion or ions to test.

$$Ag_2CO_3(s) + 2\ H_3O^+(aq) \longrightarrow 2\ Ag^+(aq) + CO_2(g) + 3\ H_2O \tag{38.7}$$

$$Ag_2SO_3(s) + 2\ H_3O^+(aq) \longrightarrow 2\ Ag^+(aq) + SO_2(g) + 3\ H_2O \tag{38.8}$$

$$Ag_3PO_4(s) + 2\ H_3O^+(aq) \longrightarrow 3\ Ag^+(aq) + H_2PO_4^-(aq) + 2\ H_2O \tag{38.9}$$

$$AgCl(s)\ or\ AgBr(s)\ or\ AgI(s) + H_3O^+(aq) \longrightarrow No\ reaction \tag{38.10}$$

The Barium Group Test

$$Ba^{2+}(aq) + SO_4^{2-}(aq) \longrightarrow BaSO_4 \text{ (s, } white\text{)} \tag{38.11}$$

$$3\ Ba^{2+}(aq) + 2\ PO_4^{3-}(aq) \longrightarrow Ba_3(PO_4)_2 \text{ (s, } white\text{)} \tag{38.12}$$

$$Ba^{2+}(aq) + SO_3^{2-}(aq) \longrightarrow BaSO_3 \text{ (s, } white\text{)} \tag{38.13}$$

$$Ba^{2+}(aq) + CO_3^{2-}(aq) \longrightarrow BaCO_3 \text{ (s, } white\text{)} \tag{38.14}$$

This test, like the silver group test, may help eliminate several anions from among the ones to be tested for individually.

To 5 drops of a test solution of the sample in water, add 3 M NH_3 solution a drop at a time until the solution is alkaline to litmus. To the alkaline solution, add 4 drops of 0.1 M barium chloride solution. If no precipitate forms, sulfate (SO_4^{2-}), phosphate (PO_4^{3-}), carbonate (CO_3^{2-}), and sulfite (SO_3^{2-}) ions are absent and need not be tested for further.

If a precipitate forms, add 6 M HCl solution until the solution is acidic; then add 5 drops more. If all the precipitate dissolves, sulfate ion is absent from the sample, but each of the other ions of the sulfate group must be tested for individually. If none of the precipitate dissolves in HCl solution, sulfate ion only, of the sulfate group ions, is present. However, it is very difficult, if not impossible, to be sure that none of a precipitate dissolves. It is safer to make individual tests for all ions of the group, even if the precipitate seems completely insoluble in acid. Table 38-1 summarizes the behavior of the various anions in reaction with specific cations.

Table 38-1. The Behavior of Various Anions with Specific Cation Reagents

Anions	Ag^+	Ba^{2+}
NO_3^-, NO_2^-	Soluble	Soluble
Cl^-, Br^-, I^-	AgCl, white; AgBr, cream; AgI, yellow. All insoluble in HNO_3. AgCl soluble, AgBr slightly soluble, and AgI insoluble, in $NH_3(aq)$	Soluble
SO_4^{2-}	Moderately soluble	$BaSO_4$, white, insoluble in HNO_3
SO_3^{2-}	Ag_2SO_3, cream-colored, soluble in HNO_3	$BaSO_3$, white, soluble in HNO_3
PO_4^{3-}	Ag_3PO_4, yellow, soluble in HNO_3	$Ba_3(PO_4)_2$, white, soluble in HNO_3
CO_3^{2-}	Ag_2CO_3, white, soluble in HNO_3	$BaCO_3$, white, soluble in HNO_3

$$BaSO_4(s) + H_3O^+(aq) \longrightarrow \text{no reaction} \tag{38.15}$$

$$Ba_3(PO_4)_2(s) + 4\ H_3O^+(aq) \longrightarrow 3\ Ba^{2+}(aq) + 2\ H_2PO_4^-(aq) + 2\ H_2O \tag{38.16}$$

$$BaSO_3(s) + 2\ H_3O^+(aq) \longrightarrow Ba^{2+}(aq) + SO_2(g) + 3\ H_2O \tag{38.17}$$

$$BaCO_3(s) + 2\ H_3O^+(aq) \longrightarrow Ba^{2+}(aq) + CO_2(g) + 3\ H_2O \tag{38.18}$$

Tests for Individual Anions

1. Test for Sulfate Ion

$$Ba^{2+}(aq) + SO_4^{2-}(aq) \longrightarrow BaSO_4(s) \tag{38.19}$$

To 2 mL of the test solution, add 6 M HCl by drops until the solution is slightly acidic. Then add 1 mL of 0.1 M $BaCl_2$ solution, or more as needed to complete the precipitation. A white precipitate of $BaSO_4$ (reaction [38.19]) confirms the presence of SO_4^{2-} ion. (Save this solution for sulfite.)

2. Test for Sulfite Ion

$$3\ H_2O + Br_2(aq) + SO_3^{2-}(aq) \longrightarrow SO_4^{2-}(aq) + 2\ Br^-(aq) + 2\ H_3O^+(aq) \tag{38.20}$$

$$SO_4^{2-}(aq) + Ba^{2+}(aq) \longrightarrow BaSO_4(s) \tag{38.21}$$

Centrifuge the solution saved from the sulfate test to obtain a clear centrifugate, add a drop or more of 0.1 M $BaCl_2$ to be sure all SO_4^{2-} is precipitated, and if necessary, add more $BaCl_2$; then recentrifuge. To the clear solution, add 1 to 2 mL of bromine water to oxidize any SO_3^{2-} to SO_4^{2-}. Stir and centrifuge. A second white precipitate of $BaSO_4$ (reaction [38.21]) now confirms the presence of SO_3^{2-} ion.

3. Test for Carbonate Ion

$$2\ H_3O^+(aq) + CO_3^{2-}(aq) \longrightarrow 3\ H_2O + CO_2(g) \tag{38.22}$$

$$CO_2(g) + Ca(OH)_2(aq) \longrightarrow CaCO_3(s) + H_2O \tag{38.23}$$

Fit a 15 cm test tube with a one-hole rubber stopper and bent delivery tube (Figure 38-1). Place about 3 mL of the test solution in this test tube. If sulfite ion (SO_3^{2-}) is present in the unknown, add 1 mL of 3% H_2O_2 to oxidize it to sulfate ion. Now insert the delivery tube into some clear limewater*, $Ca(OH)_2(aq)$, in another test tube. When ready, remove the stopper just enough to add a little 6 M HCl to the test solution. Immediately close the stopper again, and heat the tube gently to boiling to drive any CO_2 gas into the limewater. Be careful not to let any of the boiling liquid escape through the delivery tube into the limewater. A white precipitate in the limewater (reaction [38.23]) indicates the presence of CO_3^{2-} in the test solution.

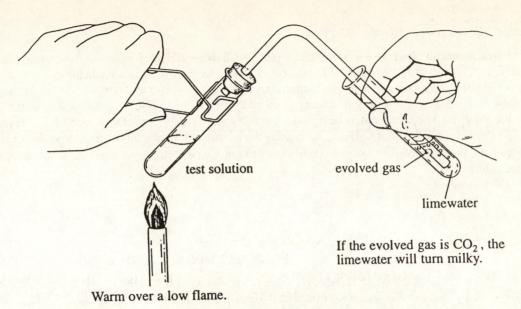

test solution   evolved gas

limewater

If the evolved gas is CO_2, the limewater will turn milky.

Warm over a low flame.

Figure 38-1. The Carbonate Ion Test

4. Test for Chloride Ion

$$Ag^+(aq) + Cl^-(aq) \longrightarrow AgCl(s) \tag{38.24}$$

$$AgCl(s) + 2\,NH_3(aq) \longrightarrow [Ag(NH_3)_2]^+(aq) + Cl^-(aq) \tag{38.25}$$

$$[Ag(NH_3)_2]^+(aq) + Cl^-(aq) + 2\,H_3O^+(aq) \longrightarrow AgCl(s) + 2\,NH_4^+(aq) + 2\,H_2O \tag{38.26}$$

To a 2 mL portion of the test solution, add a few drops of 6 M HNO_3, as needed, to make the solution slightly acid. (Test with litmus paper.) Add 1 mL of 0.1 M $AgNO_3$. (No precipitate here proves the absence of Cl^-, Br^-, or I^-.) Centrifuge the mixture. Test the clear solution with 1 drop of 0.1 M $AgNO_3$, for complete precipitation. If necessary, centrifuge again. Discard the solution. Wash the precipitate with distilled water to remove excess acid and silver ion. To this precipitate add 3 mL of distilled water, 4 drops of 6 M $NH_3(aq)$, and 0.5 mL of 0.1 M $AgNO_3$. (The proportions are important, as we wish to dissolve only the AgCl from any mixture of AgCl, AgBr, and AgI.) $[Ag(NH_3)_2]^+$ and Cl^- will form. Shake the mixture well, and centrifuge. Transfer the clear solution to a clean test tube, and acidify with 6 M HNO_3. A white precipitate of AgCl (reaction [38.26]) confirms the presence of Cl^-.

5. Test for Iodide Ion

$$2\,Fe^{3+}(aq) + 2\,I^-(aq) \longrightarrow I_2(aq) + 2\,Fe^{2+}(aq) \tag{38.27}$$

To 2 mL of the test solution, add 6 M HCl to make the solution acidic. If SO_3^{2-} or CO_3^{2-} is present, boil the solution to remove these ions. Add 1 mL of 0.1 M $FeCl_3$ to oxidize any I^- to I_2. (Br^- is not oxidized by Fe^{3+}.) Add 1 mL of trichloroethane ($C_2H_3Cl_3$), and agitate the mixture. A purple color in the $C_2H_3Cl_3$ layer confirms the presence of I^-. (Save the mixture for the Br^- test.) Verify your identification by performing this test on 2 mL of known iodide ion solution.

---

*Since $Ca(OH)_2$ (limewater) is only slightly soluble in water, its solution will appear milky. Hence the limewater should either be centrifuged or allowed to settle for 10 min to achieve clarity before being used in this test.

6. Test for Bromide Ion

$$Cl_2(aq) + 2 Br^-(aq) \longrightarrow Br_2(aq) + 2 Cl^-(aq) \qquad [38.28]$$

If no I^- was present in the above mixture, add 2 mL of chlorine water, and agitate it. An orange-brown color in the $C_2H_3Cl_3$ layer indicates Br^-. If I^- was present, separate, by means of a medicine dropper, as much as possible of the preceding iodide test solution above the $C_2H_3Cl_3$ layer that contains the I_2, and place it in a clean test tube. Again, extract any remaining I_2 by adding 1 mL of $C_2H_3Cl_3$, agitating the mixture, and separating the solution. The solution may be boiled a moment to remove any remaining trace of I_2. Then add 2 mL of chlorine water and 1 mL of $C_2H_3Cl_3$, and agitate the mixture. An orange-brown color in the $C_2H_3Cl_3$ layer indicates the presence of Br^-. Verify your identification by performing this test on 2 mL of known bromide ion solution.

7. Test for Phosphate Ion

$$3 NH_4^+(aq) + 12 MoO_4^{2-}(aq) + H_3PO_4(aq) + 21 H_3O^+(aq) \longrightarrow$$
$$(NH_4)_3PO_4{\cdot}12\ MoO_3(s) + 33\ H_2O \qquad [38.29]$$

First mix about 1 mL of 0.5 M $(NH_4)_2MoO_4$ reagent with 1 mL of 6 M HNO_3. [If a white precipitate forms, dissolve it by making the solution basic with $NH_3(aq)$, then reacidify with HNO_3.] Add 2 mL of unknown solution to this acidic molybdate ion solution. A yellow precipitate of $(NH_4)_3PO_4{\cdot}12\ MoO_3$, appearing at once or after warming a few minutes to about 40 °C, indicates the presence of PO_4^{3-} (reaction [38.29]). Verify your identification by performing this test on 2 mL of known phosphate ion solution.

8. Test for Nitrite Ion

$$2 NO_2^-(aq) + H_2SO_4(l) + 2 H_3O^+(aq) \longrightarrow 2 NO_2(g) + SO_2(g) + 4 H_2O \qquad [38.30]$$

$$NO_2^-(aq) + I^-(aq) + 2 H_3O^+(aq) \longrightarrow NO(g) + I_2(g) + 3 H_2O \qquad [38.31]$$

Evidence that nitrite ion is present will be found when concentrated sulfuric acid is added to 2 mL of the test solution or directly to the solid sample (use hood). Cold sulfuric acid produces nitrogen dioxide, NO_2, a red-brown gas, with nitrite. If no trace of brown gas is observed with 1 drop of cold, concentrated sulfuric acid, nitrite ion is not present. This test depends on the ability of nitrite ion to oxidize iodide ion to free iodine in a slightly acid solution. To confirm NO_2^-, place 3 drops of test solution in a test tube. Add 3 drops of water, 3 drops of 6 M acetic acid, 3 drops of 2 M potassium or sodium acetate, 1 drop of 0.5 M potassium iodide solution, and 4 or 5 drops of $C_2H_3Cl_3$. Shake for 10 sec; if the $C_2H_3Cl_3$ layer becomes violet in color as a result of dissolving free iodine, nitrite ion was present in the sample. Nitrate ion will not interfere by giving the same result.

9. Test for Nitrate Ion

$$NO_3^-(aq) + 3 Fe^{2+}(aq) + 4 H_3O^+(aq) \longrightarrow 3 Fe^{3+}(aq) + 6 H_2O + NO(aq) \qquad [38.32]$$

$$Fe^{2+}(aq) + NO(aq) \longrightarrow Fe(NO)^{2+}(aq) \qquad [38.33]$$

If Br^-, I^-, or CO_3^{2-} are absent from the unknown as determined previously, use 2 mL of test solution acidified with 3 M H_2SO_4. Add 1 mL of freshly prepared saturated $FeSO_4$.* Incline the test tube at about a 45° angle, and pour about 1 mL of concentrated H_2SO_4 slowly down the side of the test tube. Be careful to avoid undue mixing. A brown ring of $Fe(NO)^{2+}$ at the interface of the two liquids indicates that NO_3^- is present. A faint test may be observed more easily by holding the test tube against white paper and looking toward the light (Figure 38-2).

---

*Mix sufficient solid $FeSO_4$ with about 2 mL distilled water until some solid remains undissolved. The solution is then saturated.

If Br^- and I^- are present in the unknown, free Br_2 and I_2 may form at the interface with the concentrated H_2SO_4 and invalidate the test. If this happens, add 4 mL of a saturated solution of silver acetate to 2 mL of the test solution to precipitate AgBr or AgI. Add a drop or two of 6 M HCl to precipitate any excess silver ion as AgCl. Decant the liquid into a test tube and centrifuge. Treat the clear centrifugate with $FeSO_4$ and concentrated H_2SO_4, as directed in the previous paragraph.

If CO_3^{2-} is present, acidify the solution with 6 M H_2SO_4 until it is just acidic to litmus and then add 4 drops in excess. Place the test tube in a hot water bath until effervesence ceases. Treat the clear centrifugate with $FeSO_4$ and concentrated H_2SO_4, as directed in the nitrate test.

Please note that Ba^{2+} and NO_3^- will not both be in any unknown consisting of a soluble mixture of salts. Hence, if a precipitate is formed upon the addition of 3 M H_2SO_4 and $FeSO_4$ during the test for NO_3^-, Ba^{2+} may be present and the test solution should be tested for its presence. If Ba^{2+} is found then NO_3^- is not present and need not be tested.

After all the anions have been tested, analyze a solution of your unknown salt (or salt mixture) for cations according to the procedures in Experiments 33, 36, and 37. Include on your report any cations and anions identified.

DISPOSAL

Aqueous solutions: Dispose in a common waste bottle labeled for *qualitative analysis—solutions.*
Solids: Dispose in a waste bottle labeled *qualitative analysis—solids.*
Wash solutions/solids mixtures: Dispose in a waste bottle labeled *qualitative analysis—wash mixtures.*
All of these wastes will be collected, treated, and stored for volume reduction and proper disposal by qualified laboratory personnel.

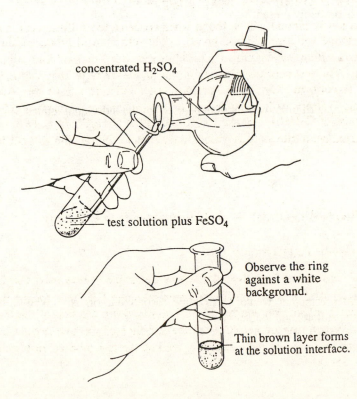

concentrated H_2SO_4

test solution plus $FeSO_4$

Observe the ring against a white background.

Thin brown layer forms at the solution interface.

Figure 38-2. The Nitrate Ion Test

Prelab   Name _____ Section _____

Analysis of Common Anions and Their Salts

1. Why do the preliminary tests not help in the identification of nitrate and nitrite?

2. A student analyzes a sample using the preliminary tests and finds that no precipitate is obtained upon the addition of silver ion but a precipitate is observed on the addition of barium ion. Which individual ion tests should be used to determine the anions present in this sample?

3. In the preliminary tests 3 M HNO_3 is added to the precipitate of the silver group. If a yellow precipitate remains, what ion is probably present?

4. Why can't a fresh sample of unknown solution be used for the individual ion test for sulfite?

5. Describe the specific tests that would need to be carried out on an unknown sample suspected of containing carbonate, sulfate, and iodide.

Appendix

I. THE INTERNATIONAL SYSTEM of Units (SI)

A. Basic SI Units

Physical Quantity	Unit	Symbol
Length	meter	m
Mass	kilogram	kg
Time	second	s
Electric current	ampere	A
Thermodynamic temperature	kelvin	K
Amount of a substance	mole	mol

B. Common Derived Units

Physical Quantity	Unit	Symbol	Definition
Electric charge	coulomb	C	$A \cdot s$
Energy	joule	J	$kg\ m^2\ s^{-1}$
Force	newton	N	$kg\ m\ s^{-1} = J\ m^{-1}$
Frequency	hertz	Hz	s^{-1}
Pressure	pascal	Pa	$kg\ m^{-1}\ s^{-2} = N\ m^{-2}$
Temperature	Celsium	°C	$273.15\ K = 0.00\ °C$
Volume	liter	L	dm^3

C. Decimal Magnitudes and Multiples

Magnitude	Prefix	Symbol	Multiple	Prefix	Symbol
10^{-1}	deci-	d	10	deca-	da
10^{-2}	centi-	c	10^2	hecto-	h
10^{-3}	milli-	m	10^3	kilo-	k
10^{-6}	micro-	μ	10^6	mega-	M
10^{-9}	nano-	n	10^9	giga-	G
10^{-12}	pico-	p	10^{12}	tera-	T
10^{-15}	femto-	f			
10^{-18}	atto-	a			

Appendix

II. FUNDAMENTAL CONSTANTS AND CONVERSIONS

Description	Symbol	Value
Avogadro's constant	N_A	6.02252×10^{23} mol^{-1}
Faraday constant	F	94687.0 C mol^{-1}
Charge of electron	e	1.60210×10^{-19} C
Mass of electron	m_e	9.1091×10^{-31} kg
Mass of proton	m_p	1.67252×10^{-27} kg
Mass of neutron	m_n	1.67482×10^{-27} kg
Planck's constant	h	6.6256×10^{-34} J s
Speed of light in vacuum	c	2.997925×10^{8} m s^{-1}
Gas constant	R	8.3143 J mol^{-1} K^{-1}
		1.9871 cal mol^{-1} K^{-1}
		82.053 cm^3 atm mol^{-1} K^{-1}
		0.082053 L atm mol^{-1} K^{-1}
Boltzmann constant	k	1.38054×10^{-23} J K^{-1}
Volume of 1 mol of ideal gas	V_m	
at 1 atm, 0 °C		22.4136 L mol^{-1}
at 1 atm, 25 °C		24.4650 L mol^{-1}
Base of natural logarithms	e	2.71828
	$\ln x$	$2.3026 \log x$
	π	3.14159265

III. COMMON CONVERSION FACTORS

Joule = Newton·meter = volt·coulomb
    = 10^7 erg
Watt = Joule s^{-1}
°F = 32 + 1.8 °C = 32 + (9/5)°C
°C = (°F − 32)/1.8 = (°F − 32)(5/9)
1 atm = 760 mmHg = 760 torr
    = 14.7 lb in^{-2}
    = 101,325 Pa = 1.013 bar
1 Angstom = 1 Å = 10^{-10} m = 10^{-8} cm
1 calorie = 4.184 J (exactly)
2.54 cm = 1 inch (exactly)
1 m = 39.37 inch
453.6 g = 1 lb
1 kg = 2.205 lb
0.946 L = 1 qt
1 L = 1.057 qt
eV = electron volt
1 eV atom^{-1} = 1.602×10^{-19} J atom^{-1}
       = 23.06 kcal mol^{-1}
       = 96.48 kJ mol^{-1}

Qualitative Analysis Report Form　　　　　　　　Name _____ Section _____

Substance	Reagent	Result	Conclusion

Appendix

Qualitative Analysis Report Form   Name _____ Section _____

Substance	Reagent	Result	Conclusion

Qualitative Analysis Report Form　　　　　　Name ＿＿＿＿＿＿＿ Section ＿＿＿＿＿

Substance	Reagent	Result	Conclusion

Qualitative Analysis Report Form   Name _____ Section _____

Substance	Reagent	Result	Conclusion

Qualitative Analysis Report Form   Name _____ Section _____

Substance	Reagent	Result	Conclusion

Qualitative Analysis Report Form   Name _____ Section _____

Substance	Reagent	Result	Conclusion

Qualitative Analysis Report Form   Name _____ Section _____

Substance	Reagent	Result	Conclusion